Chalcogenides - Preparation and Applications

Edited by Dhanasekaran Vikraman

Published in London, United Kingdom

Chalcogenides - Preparation and Applications
http://dx.doi.org/10.5772/intechopen.98001
Edited by Dhanasekaran Vikraman

Contributors
Ho Soonmin, Immanuel Kodaikanal, Mohanraj Kumar, Rakesh K. Sonker, Pronoy Nandi, Mohamed Amine Ghebouli, Brahim Ghebouli, A. Kryuchyn, V.V. Petrov, V.M. Rubish, M.L. Trunov, Devappa S. Lamani, Akrajas Ali Umar, Nurul Ain Abd Malek, Nabilah Alias, Abang Annuar Ehsan, Pramoda K. Nayak, Nilanjan Basu, Vishal K. Pathak, Laxman Gilua, Madhavi Jonnalagadda, Varishetty Madhu Mohan, VishnuBhotla Prasad, Prince Sharma, Veerpal Singh Awana, Mahesh Kumar

First published in London, United Kingdom, 2022 by IntechOpen
IntechOpen is the global imprint of INTECHOPEN LIMITED, registered in England and Wales, registration number: 11086078, 5 Princes Gate Court, London, SW7 2QJ, United Kingdom

British Library Cataloguing-in-Publication Data
A catalogue record for this book is available from the British Library

Additional hard and PDF copies can be obtained from orders@intechopen.com

Chalcogenides - Preparation and Applications
Edited by Dhanasekaran Vikraman
p. cm.
Print ISBN 978-1-80355-660-4
Online ISBN 978-1-80355-661-1
eBook (PDF) ISBN 978-1-80355-662-8

Meet the editor

Dr. Dhanasekaran Vikraman is an assistant professor in the Division of Electronics and Electrical Engineering, Dongguk University, Seoul, Korea. He received his bachelor's degree from Mannai Rajagopalaswamy Government Arts College, Mannargudi, affiliated with Bharathidasan University, India. He completed his master's degree and Ph.D. at the Department of Physics, Alagappa University, India. Later, he received a visiting scientist position at KIST, Korea; a Marie-Curie Experienced Researcher fellowship at the Department of Physics, Aristotle University of Thessaloniki, Greece; and post-doc positions at Sejong University and Ajou University, Korea. He has authored more than 180 international journal articles and 3 book chapters and edited several books. He serves as a guest editor and topic editor for various journals.

Contents

Preface

Chalcogenides have attracted the attention of the research community for decades because of their distinctive characteristics. In recent years, chalcogen-based materials have been used in various micro-electronics and energy devices, owing to their high feasibility, good mechanical properties, tunable electronic structures, and attractive electrical conductivity. This book describes the synthesis of chalcogen-based materials and their various uses. It is divided into three sections: "Chalcogen Fabrication", "Solar Cells and Batteries" and "Nanoscale Devices".

Section 1 includes four chapters that discuss the preparation of chalcogens for different applications. Chapter 1 describes the recent developments of multifunctional characteristics of chalcogen thin films. It also provides a detailed overview of the capability of chalcogen-based materials in supercapacitors, perovskite solar cells, thin-film sensors, and tandem solar cells. Chapter 2 designs the physical properties calculation of beryllium selenide (BeSe) by M.A. Ghebouli and B. Ghebouli. First-principles calculations were carried out using Cambridge Serial Total Energy Package (CASTEP) code. The authors estimate the hybrid functional parameters such as structural, elastic, electronics, and optical properties of zinc-blende and NiAs phase of BeSe. Chapter 3 summarizes spectroscopic studies of mechanically exfoliated PtSe2. It describes the electronic band structure and density of states for monolayer, bi-layer, tri-layer, and bulk PtSe2. Theoretical approximations determine the structural behavior of 1T phase PtSe2. Chapter 4 describes a series of synthetic organoselenium compounds such as ebselen analogues, diaryl selenides, spirodioxyselenurane, and spirodiazaselenuranes and their glutathione peroxidase (GPx) catalytic activity.

Section 2 includes two chapters on energy-based applications of chalcogenides such as perovskite solar cells and batteries. Chapter 5 describes the fabrication of perovskite solar cells using transition metal dichalcogenide (TMDs) as the electron transport layer. Moreover, it investigates the role of TiS2-, MoS2-, and WS2-based electron transport layers in perovskite solar cells. Chapter 6 discusses chalcogen (sulfur, selenium, and tellurium (SSTs))–tuned batteries. It describes aluminium ion batteries supported by SST cathodes. It also demonstrates various chalcogen substitutions in Li-ion and metal-ion batteries.

Section 3 includes chapters related to the fabrication of semiconducting nanodevices using chalcogen materials. Chapter 7 discusses the temperature-dependent assessment of conductivity, terahertz generation, and charge carrier dynamics. It also discusses chalcogenides such as bismuth selenide, bismuth telluride, and antimony telluride and their charge movements. Finally, Chapter 8 describes the recording characteristics of chalcogenides.

This edited volume discusses the preparation of chalcogenides and their circular angle uses in different industries.

Dhanasekaran Vikraman
Division of Electronics and Electrical Engineering,
Dongguk University-Seoul,
Seoul, Korea

Section 1

Chalcogen Fabrication

Chapter 1

Recent Developments on the Properties of Chalcogenide Thin Films

Ho Soonmin, Immanuel Paulraj, Mohanraj Kumar, Rakesh K. Sonker and Pronoy Nandi

Abstract

Chalcogenide thin films have attracted a great deal of attention for decades because of their unique properties. The recent developments on thin film-based supercapacitor applications were reported. As a result of sustained efforts, the experimental findings revealed remarkable properties with enhanced fabrication methods. The properties of perovskite solar cells were discussed in terms of crystal structure and phase transition, electronic structure, optical properties, and electrical properties. Perovskite solar cell has gained attention due to its high absorption coefficient with a sharp absorption edge, high photoluminescence quantum yield, long charge carrier diffusion lengths, large mobility, high defect tolerance, and low surface recombination velocity. The thin film-based gas sensors are used for equally the identification and quantification of gases, and hence should be both selective and sensitive to a required target gas in a mixture of gases. Metal chalcogenide materials are considered excellent absorber materials in photovoltaic cell applications. These materials exhibited excellent absorption coefficient and suitable band gap value to absorb the maximum number of photons from sun radiation. The photovoltaic parameters were strongly dependent on various experimental conditions.

Keywords: thin films, semiconductor, band gap, solar cells, sensor, chalcogenide materials, supercapacitor, power conversion efficiency

1. Introduction

Chalcogenide thin films have attracted a great deal of attention for decades because of several reasons such as earth abundancy, environmental friendly [1–3], excellent structural [4–6], electrical [7–9], and optical properties [10–12]. These materials could be employed in various applications such as solar cells [13–15], ultraviolet light emitters, laser devices [16], spin functional devices, gas sensors, transparent electronics, corrosion resistant coating [17], microelectronics [18], optics, magnetic and acoustic wave devices. Several deposition methods including chemical vapor deposition [19], physical vapor deposition [20], sputtering [21], SILAR, spray pyrolysis [22], chemical bath deposition [23–26], vacuum evaporation method [27], and electro deposition method [28, 29] have been used to produce thin films. Generally, these deposition methods could be divided into physical method and chemical techniques. Researchers highlighted that the chemical deposition method has many advantages such as inexpensive, and convenient for large area deposition [30–32].

Supercapacitor has been considered as one of the potential energy storage systems. The redox electrochemical capacitors and the electrochemical double layer capacitor have been extensively investigated by many researchers. Researcher reported that transition metal oxide, conducting polymers, and metal oxide thin films have been tested in supercapacitors. The perovskite structure consisted of the crystal structure of calcium titanium oxide. This material showed high absorption coefficient with a sharp absorption edge. The organic-inorganic hybrid perovskites based solar cell was made from sandwiching a perovskite absorber layer between the electron transport layer and hole transport layer. The obtained solar cells showed some unique advantages such as low-temperature processes for all sub cells, compatibility with flexible and lightweight applications. The thin film-based sensor could be used to convert physical or chemical quantity into equivalent electrical for measurement. Sensor is critical in improving the reliability and efficiency of manufacturing operations by providing faster and more accurate feedback regarding product quality.

In this book chapter, thin film based solar cell, thin film based supercapacitor and thin film based sensor will be discussed. The properties of the obtained films were reported. Lastly, the advantages and limitations of these materials will be highlighted.

2. Thin film based supercapacitors

Recently electronic devices such as computers, roll-up displays smartwatches, mobile phones and other portable devices abound in the twenty-first century. For greater performance, improved energy storage devices are required to reduce the energy consumption of these smart electronic devices [33]. As a result, devices with long-lasting battery, high power outputs, and quick recharge times are required. As a consequence, it is critical to create innovative energy storage materials and devices. The realities of scarcity of fossil fuels, and environmental damage should all be considered in this endeavor [34]. By modify the surface properties of the electrodes with a long life cycle, the supercapacitor (SC) is such an effective energy-saving technology that is environmentally friendly with quick charging, and high energy density are just a few of the benefits [35]. However, this redeemer (supercapacitor) has issues. Nevertheless, in comparison to lithium batteries, such savior (supercapacitor) has challenges such as poorer energy density, unavailability, and the high cost of ruthenium (IV) oxide (RuO_2) and platinum electrode materials, all of whom have stymied the supercapacitor development. Supercapacitors, which are versatile, compact, ecologically benign, and yet still economical energy storage devices, are in growing market. The flexible supercapacitors, which bridges the gap between batteries and traditional capacitors, is a bright spot in the realm of energy-saving engineering. Flexible-all-solid-state thin film supercapacitor, an innovative novel thing, has gotten a lot of interest as unique energy storage devices because of its friendly construction, compact size, easy handling, and excellent power density with a quick charging-discharging rate. The supercapacitor is called as electrochemical capacitors it has a fast charging and discharging properties, excellent power density and high specific capacitance with compact construction, and inexpensive cost of maintenance. The three primary mechanisms of supercapacitor can be classified (**Figure 1**), which is

DOI: http://dx.doi.org/10.5772/intechopen.102429

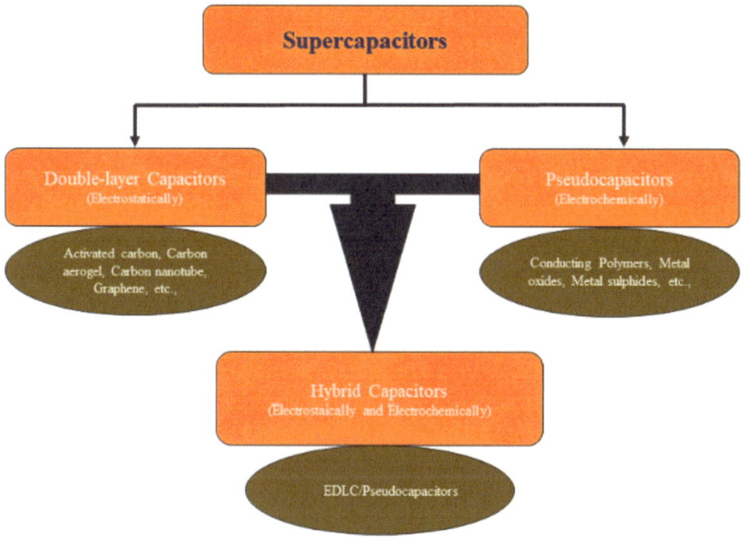

Figure 1.
Classification of supercapacitors.

depending of the reversible redox reactions and the accumulation of charge. There is electric double layer capacitors (EDLC), pseudo capacitor, and the combination of EDLC and pseudo capacitor called the hybrid supercapacitor [36].

Thin films are very intriguing in modern research for a variety of applications in ethanol sensor, photocatalytic, thermoelectric and supercapacitor [37–40]. The supercapacitors can store the electrical energy for all the electronic devices to stabilize the power supply. Generally, to prepare a pseudo capacitive electrode transition metal oxide (TMO) is the most popular approach, however relatively higher electrical resistivity restricts whose use several fields. As a consequence, the focus of researchers is turning to metal chalcogenides, which have a lower electrical resistivity than oxygen due to sulfur's low electronegativity. The preponderance of these metal chalcogenides, mostly sulfides, are made from inexpensive and abundant transition metals. For example, Dai and co-workers [41] have prepared hierarchically structured Ni_3S_2 and multi-walled carbon nanotube (MWCNT) composites using the hydrothermal methods and the prepared device can have obtained the maximum C_s of 55.8 F g^{-1}, it provides a highest energy density of 19.8 Wh/kg at power density of 789 W/kg. Xiao and co-workers [42] prepared a nickel cobalt sulfide nanoparticle graphene-based sheet ($NiCo_2S_4$@GR) there is no surfactant through simple one-step solvo thermal method, which results revealed the maximum C_s of 1708 F g^{-1} at a current density of 1.0 A g^{-1}, while comparing without graphene. Mukkabla and co-workers [43] reported a Poly(3,4-ethylenedioxypyrrole) (PEDOP) Enwrapped bismuth sulfide (Bi_2S_3) nano flowers hybrid flexible SC_S, and composite offered a maximum C_s of 329 F g^{-1} at 0.4 A g^{-1}. Furthermore, these are usually undergoing redox reactions between the metallic ions valence states. Besides, TMO and transition metal chalcogenides, various metal nitrides have previously been observed has outstanding results as electrodes in supercapacitors and lithium ion batteries with impressive results. Recently, metal nitrates also have superior abilities in electrochemical properties with excellent chemical stability. Metal nitrides have gotten a lot of interest as supercapacitors electrodes since they have a lot of benefits. Metal nitrates have three major advantages. (1) It has a high σ (electrical conductivity) of 55,500 S/cm^{-1} while compared to the metal oxides as a result shows the excellent power density, (2) compared to the metal oxides and carbon based materials metal nitrates have a higher specific capacitance, which results shows the higher energy density, and (3) high mechanical stability. These characteristics make them extremely promising as high-performance supercapacitor electrodes. Balogun and co-workers [44] have summarized the performance of different metal nitrides like molybdenum nitrides (MoN), nickel nitride, titanium nitride. Among these metal nitrides, molybdenum nitride was considered as the first metal nitride which could be used as supercapacitor electrode materials. However, for supercapacitor applications, researchers mostly considering their materials cost and electrochemical performance. There are many transition metals and metal oxides are considerable for supercapacitor applications such as CuO, NiO, Mn_3O_4, Co_3O_4, Ni or $CuCo_2O_4$ and Ni or $CuFe_2O_4$ [45–50]. Compared to the other metal oxides, the metal ferrite based materials much attracted to the researchers. For example, Fe, Ni or Cu based Fe_2O_4 materials have an excellent performance in the energy storage applications. There are two major methods could be used to prepare the thin films supercapacitors, namely physical technique (physical vapor deposition and sputtering) and chemical method. The successive ionic layer adsorption and reaction (SILAR), spin coater, and chemical bath deposition (CBD) are some examples for chemical deposition method (**Figure 2**).

Bandgar and co-workers [51] studied the nature of starting materials on the properties of $NiFe_2O_4$ thin films for flexible supercapacitors. There are several morphologies could be observed (nanosheet, flower, and feather) through different salts such as nickel(II) chloride hexahydrate ($NiCl_2{\cdot}6H_2O$), nickel nitrate [$Ni(NO_3)_2{\cdot}6H_2O$], and nickel sulfate hexahydrate ($NiSO_4{\cdot}6H_2O$), respectively. The nanosheet based electrode material received the maximum C_s of 1139 Fg^{-1}, nanoflower and feather achieved the good C_s of 677 and 435 $F\,g^{-1}$, respectively. Immanuel and co-workers [52] have

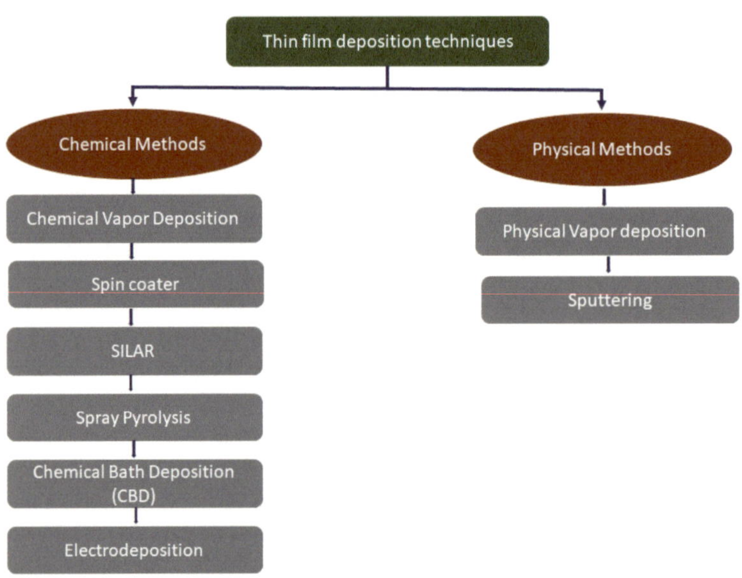

Figure 2.
Thin films deposition techniques.

optimized the Cr doped Mn_3O_4 thin films for high performance supercapacitors using the SILAR method. The experimental results showed that 3 wt % of Cr doped Mn_3O_4 thin films exhibited the maximum C_s of 181 Fg^{-1} at the current density of 1 Ag^{-1}.

Jesuraj and co-workers [53] studied the pristine and Li doped NiO thin films using the spin coating method. Kin and co-workers [54] prepared the carbon based flexible supercapacitors using the chemical vapor deposition. Yu and co-workers [55] have prepared the cobalt nickel oxide and sulfide heterostructure thin films through electrodeposition method for supercapacitor applications. The obtained findings revealed the maximum energy density of 78.2 $Wh{\cdot}kg^{-1}$ at 542.8 $W{\cdot}kg^{-1}$ and the high power density of 5440.2 $W{\cdot}kg^{-1}$. Recently, Immanuel and co-workers [56] synthesized Mn_3O_4 nanorod thin films via SILAR method. The prepared Mn_3O_4 thin films showed the maximum C_s value of 295 Fg^{-1} at the scan rate of 2 mVs^{-1}. Vivek and co-workers [57] prepared a reliable electrode material, and results obtained a maximum C_s of 426.40 Fg^{-1} at a current density of 1 Ag^{-1}. Arulraj and co-workers [58] prepared the cubic shaped Ag_2S using the CBD method on Ni mesh. The prepared Ag_2S used a working electrode, which electrochemical performance showed the highest C_s of 179 C/g at constant charge and discharge current density of 1 A/g.

3. Thin film based perovskite solar cells

Any materials which have the crystal structure of calcium titanium oxide ($CaTiO_3$), were known as the perovskite structure and the materials have stoichiometry of ABX_3; where "A" is the larger cation, "B" is the smaller cation and "X" is the anion. Each unit cell of ABX_3 crystal comprises of corner sharing BX_6 octahedra, with the "A" moiety cubo-octahedral cavity. In case of organic-inorganic hybrid perovskites (OIHP), halide anions (I^-, Br^-, Cl^-) are found at the "X"-site anion instead of oxygen, while monovalent ($CH_3NH_3^+$, $CH(NH_2)_2^+$) and bivalent (Pb^{2+}, Sn^{2+}) cations occupy the "A" and "B" sites, respectively. Halide perovskites were first reported by Moller in 1958 for cesium lead halides [59]. Further, it was also observed that small organic molecules with effective radii less than 260 pm [methylammonium (MA), formamidinium (FA), hydrazinium, hydroxylammonium) can also accommodate inside the PbX_6 octahedrons. The word "hybrid" indicates that the crystal is made specifically by the combination of "organic" and "inorganic" components. The architecture of OIHP-based solar cell is quite simple and prepared by sandwiching a perovskite absorber layer between the electron transport layer (ETL) and hole transport layer (HTL). A standard OIHP based solar cell device has a structure composed of glass/ transparent conductive oxide (TCO)/TiO_2 (ETL)/ mesoporous TiO_2 (mp-TiO_2)/ perovskite (~500 nm)/ HTL/ metal and a quite high efficiency exceeding 20% can be realized without including complicated processing steps. The operation of the perovskite device is sstraight forward; namely, the photo-electrons and holes created by light absorption are collected in the ETL and HTL, respectively, and the electrons flow through the outer circuit and recombine with holes at the HTL/metal interface. The efficiencies of OIHP-based solar cells have increased all the way from 3.8% in 2009 to 25.5% for single-junction solar cells, and 29.15% for the highest publicly disclosed perovskite/silicon (Si) tandem [60].

The properties of perovskite solar cells were discussed in terms of crystal structure and phase transition, electronic structure, optical properties and electrical properties. One of the interesting aspect of the crystal structure of halide perovskite is the structural flexibility of organic cation. Taking $MAPbI_3$ as an example, the disorder-order transition of MA^+ cation is believed to trigger the phase transition with the decrease of temperature. At high temperature $MAPbI_3$ takes a cubic structure (space group: Pm-3 m; Z = 1). Since MA^+ has a lower symmetry of C_{3v}, the orientation of MA^+ ion should be disordered to satisfy the O_h symmetry. As the temperature is lowered, tetragonal and orthorhombic phases are realized by an accompanying ordering of methylammonium ion. Structural transition from cubic to tetragonal phase occurs due to the reorientation of MA^+ ion, as observed by nuclear magnetic resonance (NMR) studies where lowering the number of disorder states of MA^+ was observed from 24 in the cubic phase to 8 in tetragonal phase [61]. Below a critical temperature (tetragonal-orthorhombic phase transition), the MA^+ molecule is frozen (only 1 degree of freedom) and the symmetry of $MAPbI_3$ become orthorhombic. Similar crystallographic phase transition can be realized with replacing I by Br and Cl [62].

The band structure of $MAPbI_3$ exhibits a direct bandgap of 1.6 eV at the R point. Calculated band structure suggests conduction band minima (CBM) is dominated by the Pb-*p* orbital, whereas the valence band maxima (VBM) is constituted by I-*p* states mixed with a small amount of Pb-*s* states, which is consistent with the photoemission results [63]. The optical transition of $MAPbI_3$ relies on a direct bandgap *p-p* transition, leading to a strong optical absorption coefficient. Strong *s-p* antibonding enhances dispersion of the upper valence bands [64], which resulted in small effective masses of electrons (m_e^*) and holes (m_h^*). Further, it is believed that the role of MA^+ cation is to maintain the overall charge symmetry and as dictated by the crystal structure of the system [65]. However, it was reported that MA^+ cation has an indirect impact on the shape and orbital composition of the band edges. The molecular orientation of MA^+ cation can distort the PbI_6 octahedral and affected the cell size and bonding of Pb-I, which modulated the density of states near the band edges [66]. Other halide perovskites also possess similar ways of electronic band structure.

Organic-inorganic hybrid perovskites are direct band gap semiconductor and the direct transition produces large absorption coefficients of the order of 10^4–10^5 cm^{-1}. In the case of perovskite thin films, the optical properties of perovskites are dramatically affected by the quality, composition and morphology of the film [67]. Sizes of the halide anions (X = I, Br, Cl) affected the electronic band structure of the system. Large anion (iodine based materials) showed a smaller bandgap and corresponded the absorption edge at 780 nm; whereas substituting iodine with smaller bromine (chlorine) anion shifts the absorption edge to 535 nm (408 nm) for MA^+ based perovskite system [68]. A systematic blue shift of the PL emission peak is observed with the increase of Br concentration in mixed halide perovskite of the type $MAPb(I_{1-x}Br_x)_3$. Further, replacing MA with $CH(NH_2)_2$ red shifts the absorption spectra by 40 nm, which makes $CH(NH_2)_2PbI_3$ more suitable for high-performance solar cell applications [69]. Intermediated solid solutions of $MASn_{1-x}Pb_xI_3$ with x = 0.25 and 0.5 exhibited the smallest band gap of 1.17 eV [70]. Irrespective of bandgap tuning, fundamental understanding of absorption and PL spectra are essential to study the basic photo physical properties of hybrid perovskite. In spite of several optical investigations performed at different temperatures, there have been a lot of ambiguities in the data as well as its interpretations, especially observation of multiple peaks in the photoluminescence (PL) spectrum of organic-inorganic hybrid perovskites. Literature reports excitonic emission, tetragonal inclusion in orthorhombic phase, order-disorder transition, surface-bulk effects are responsible for these multiple PL emissions [71].

Space charge limited current (SCLC) is one of the effective approaches to measure mobility, diffusion length and trap density of hybrid perovskites. Due to the advancement in fabrication techniques, the diffusion length of hybrid perovskite has increased from 1 to ~10 μm in about 3 years [72]. This improvement reflects the progress that has been recently made in producing samples with better structural order and morphology. Further, it is also observed that the diffusion length has a strong dependence on the grain size of the film. The results showed that more than 1 μm diffusion length has been achieved by realizing films with an average grain size of 2 μm. The perovskite single crystal was found the highest measured diffusion length (10 μm) [73]. Carrier mobility of hybrid perovskite has also been improved over the years and exhibited morphology dependence. Mobility values exceeding 10 $cm^2V^{-1}s^{-1}$ have been measured in perovskite film [74] and above 100 $cm^2V^{-1}s^{-1}$ in perovskite single crystals. Further, it is also observed that the mobility (and also diffusion length) did not exhibit a strong dependence on the material composition. Further, the dielectric constant (relative permittivity) is a complex number given by, $\varepsilon = \varepsilon^{/} - \varepsilon^{//}$, where the real part $\varepsilon^{/}$ is the charge storage ability and the imaginary part $\varepsilon^{//}$ is the energyloss. For $MAPbI_3$, a small $\varepsilon^{/}$ is obtained ($\varepsilon^{/}$ = 6.5 in experiment, while 5.6 to 6.5 in calculation) at optical frequency and only electronic polarization takes part in dielectric process [75]. With the decrease of frequency, ionic polarization and dipolar polarization (contribution from MA^+ dipoles) leading to enhanced $\varepsilon^{/}$ ($\varepsilon^{/}_{low}$ ~ 60 at 100 KHz).This large dielectric constant facilitates the screening effect of Coulombic attraction between photoexcited electron-hole pairs (excitons), so that they can be separated easily. Also, noncentro symmetric crystal structure in tetragonal and orthorhombic phases proposed OIHP are ferroelectric in nature. It is also believed that ferroelectricity may give rise to hysteresis observed in current-voltage (I-V) curves. However, observation of ferroelectricity in hybrid perovskite is not well justified from polarization-electric field (P-E) hysteresis loop and second harmonic generation experiments. Despite the above controversies, it is of great interest to study the order-disorder transition of hybrid perovskites due to MA^+ orientation inside the PbX_6 octahedral [76].

Perovskite solar cell has gained attention due to favorable material properties of OIHP, which include a high absorption coefficient with a sharp absorption edge, high photoluminescence quantum yield, long charge carrier diffusion lengths, large mobility, high defect tolerance, and low surface recombination velocity. At the same time, easy solution processability and completely tunable optical bandgap from blue to red regions of wavelength just by mixing the B-site cation (Pb-Sn) and the X-site anion (I-Br-Cl), while maintaining the sharp absorption edge makes the OIHP family a potential candidate for application in multijunction/tandem solar cells. Another strong advantage of hybrid perovskite solar cells is quite high V_{oc}, which can be explained by the suppression of the defect formation in the bulk layer as well as at the interfaces. It is observed that OIHP solar cells with E_g < 1.65 eV, the open circuit current (V_{oc}) is remarkably high and V_{oc} increases with band gap (E_g) without significant V_{oc} loss. In particular, a quite high V_{oc} of 1.26 V has been reported for a pure $MAPbI_3$ cell, which is very close to the theoretical limit of 1.32 V, with V_{oc} loss of only 60 mV. High absorption coefficient and low nonradiative recombination rate of OIHP solar cell resulted very small short circuit current (J_{sc}) loss. We know that photovoltaic devices rarely operate at room temperature, and high power output is necessary even under high-temperature operation conditions. It is observed that OIHP solar cell shows the lowest temperature coefficient (TC) of −0.17%, which is far better than other photovoltaic devices. Also nearly 90% efficiency is maintained even at a high operating temperature of 85°C. Further, OIHP-based solar cell exhibited some unique advantages such as low-temperature processes for all sub cells, compatibility with flexible and lightweight applications, low life-cycle environmental impacts and embodied energy, and potentially low fabrication costs.

Although OIHP solar cells produced quite impressive efficiency, they have several limitations too and to overcome these limitations are the major challenge for the commercialization of these devices. One significant drawback of OIHP is degradation of these perovskite materials under a range of environmental factors such as humidity, illumination, oxygen, and thermal stress. OIHP solar cells are ionic crystals, and the presence of H_2O leads to the decomposition of the perovskite structures to hexagonal-shaped PbI_2/hydrate crystals; which can be suppressed by introducing protective (passivation) layers. In case of mixed halide perovskites strong photo-induced phase segregation occurred under illumination and judicial choice of A-site cation can minimize this instability. Further, it is observed that a higher level of performance in OIHP solar cell is hindered by anomalous hysteretic behavior and large discrepancy between the forward and reverse scans put a question on the reproducibility of power conversion efficiency (PCE) of the device. In searching for the possible origins of hysteresis, several explanations such as ion migration, charge trapping/detrapping, photoinduced capacitive effect, and ferroelectricity have been imposed. Among them, ion migration and ferroelectricity are believed as feasible origins of the hysteresis in transport measurements. Extensive research efforts continue to find the long-term stability of OIHP solar cells.

Another major challenge is the realization of large-area module due to its fabrication limitations. Till now high efficiency of 17.9% has been realized for the large-area module with a size of 30 × 30 cm^2 (aperture area: 80^2 cm^2), which was formed by an inkjet printing technology. Thus development of proper fabrication technique is essential to make pinhole free large-area OIHP devices. Also in the large area tandem cells, current matching conditions for the top and bottom cells as well as each sub cell need to be established; which can be improved through technological advances.

High toxicity of heavy metal (lead) is a serious problem which cannot be neglected in OIHP-based solar cells. Although the content of lead (Pb) in OIHP solar panel (~1m^2) is only a few hundred milligrams, could be severe problems in environmental impact. As an alternative people are trying to replace Pb^{2+} with Sn^{2+}; but the efficiency of Sn-based photovoltaic devices are extremely poor. Thus, roof-top application of OIHP modules is difficult and large-area operations as solar farms are more appropriate. Also, encapsulation of photovoltaic module and environmentally friendly 100% recycling programs are essential for OIHP-based solar modules.

The future of perovskite solar cells was highlighted. As discussed earlier, the significantly reduced efficiency upon solar module area scaling-up is still the main challenge to face for the commercialization of OIHP-based solar cell. It is observed that efficiency decreases to 19.6% when the aperture area increases from 0.1 cm^2 to about 10 cm^2, and further drops to 17.9% with the area approaching 1000 cm^2, which still lags far behind that of the crystalline silicon cells (26.7% at 79 cm^2 and 24.4% at 13,177 cm^2). Thus, intensive works should be conducted to precisely control the uniformity of the crystallization process in large-area perovskite films. Also, the fundamental photophysical mechanisms relative to the efficiency loss in OIHP modules should be further studied to understand role of surface and interface. Development of green solvent systems or the solvent-free deposition technology for fabricating large-area perovskite film will be an important research topic in the future. Besides the efficiency, more and more attention need to invest in the long-term stability of OIHP solar modules. Recently, Okinawa Institute of Science and Technology Graduate University in Japan reported over 1100-h operational lifetime for a 10 × 10 cm^2 solar module. Although many research groups and companies claimed that their devices have passed International Electro Technical Commission (IEC) standard test, there are still some stability issues needed to be addressed at the next stage. Thus proper development of encapsulation technology is essential and we believe that a growing number of studies will move to exploit such multifunctional encapsulation materials in the near future. The single-junction OIHP cells with efficiency above 24% and long-term stability can be more cost-effective than tandem cells which may work at a PCE of 27–28%. Thus, more efforts should be made in fabrication and scaling up of single-junction OIHP-based solar cells with high efficiency, high yield, and long-term stability. Development of low-cost large-scale fabrication methods with highly reproducible results is required for commercialization of OIHP-based photovoltaic cells.

4. Thin film based sensor

A thin film-based sensor is a type of transducer which converts a physical or chemical quantity into equivalent electrical for measurement. It is used to detect the presence of stimulus to very low concentrations of toxic or harmful target environment (gases) of importance, such as ammonia [77], carbon monoxide [78], carbon dioxide [79], nitrogen dioxide [80], sulfur dioxide [81], propane [82], liquefied petroleum gas [83], hydrogen sulfide [84], and volatile organic compounds. Worldwide thin film gas sensing technology is playing a major role in protecting the environment and improving homeland security. Sensors are also critical in improving the reliability and efficiency of manufacturing operations by providing faster and more accurate feedback regarding product quality. In the area of environmental health and safety, lowering the limits of detection can improve the quality of life through precise information regarding the pollutants in air, water and soil. High-performance thin-film sensors and systems are essential to monitoring various kinds and quantities of analysts.

The typically thin film-based sensors are described using the main characteristics such as sensing response, stability, repeatability, reproducibility, linearity, response time, and recovery time. An efficient thin-film sensor;

- Must have a high sensing response towards a very low concentration of target gas.
- Would give the same sensing characteristics after repeated usage (stability) and for different sensors of the same kind.
- Should be capable of responding fast towards a target gas.
- Must regain initial characteristics as soon as the target gas is flushed.
- Thin-film sensor response should increase linearly with increasing the concentration of target gas.

The thin film-based gas sensors are used for equally the identification and quantification of gases, and hence should be both selective and sensitive to a required target gas in a mixture of gases. Sensitivity defines the smallest concentration of gas/vapour that can be fruitfully and repeatedly sensed by a thin film sensor.

Thin film-based semiconductor is commonly used materials as sensor application as indicated in **Table 1**. This is because of its versatile advantages like high sensitivity and low manufacturing is metal oxide which contain the elements having one oxidation state because it requires more energy to form more than one oxidation states.

S. No.	Material/modifier	Temperature (°C)	Gas concentration (ppm)	Response	Response/ recovery time	Reference
1.	SnO_2/Pt	200	1000	89	20/27 sec	[85]
2.	ZnO/PANI	36	1000	13	3.3/9.8 min	[86]
3.	TiO_2/PANI	273	0.1 vol%	0.63	3.3/3.0 min	[87]
4.	TiO_2/Ni	250	1000	37	–/–	[88]
5.	ZnO/PEDOT: PSS	27	1000	0.58	3.7/3.1 min	[89]
6.	ZnO/MWCNT	30	1500	61	5.8/3 min	[90]
7.	h-BN/–	RT	3.0 vol%	6.17	55/40 sec	[91]
8.	PANI/–	RT	100	12.10	11/07 sec	[92]
9.	ZnO-TiO_2/PANI	30	20	412	35/54 sec	[93]
10.	CdS/–	70	20	173	5.52/3.46 min	[94]
11.	Ag-$BaTiO_3$/CuO	250	5000	0.28	15/10 min	[95]
12.	CuO-$Cu_xFe_{3-x}O_4$	250	3000	0.50	9.5/– min	[96]
13.	CdO	250	5000	0.01	3.33/5 min	[97]
14.	PEDOT-BPEI	RT	1000	0.03	–/60 min	[98]
15.	La1_xSrxFeO3	380	2000	0.25	11/15 min	[99]
16.	ZnO	200	3000	0.03	8 /40 sec	[100]
17.	ZnO-La (50%)	400	5000	0.65	90/38 sec	[101]
18.	SnO_2/PANI/Ag	30	1000	67	1000/900 sec	[102]
19.	TiO_2/Zn	RT	1.5 vol%	2.92	120/– Sec	[103]
20.	Fe_2O_3/ PANI	RT	20	229	2.35/3.8 min	[104]

Table 1.
Literature survey of various gas sensing characteristics of different metal oxidebased nanomaterials with different modifiers.

Semiconductor metal oxide films have been exploited for the sensing of various toxic and harmful gases in the form of ceramics, thick films, thin films or nanostructures. Sensors based on ceramics have shown advantages in terms of their mechanical strength, large resistance to chemical attack and good thermal and physical stability and most of the available commercial sensors are based on ceramics only. One of the additional attractive features associated with low temperature operated semiconductor thin film sensor is that it can lead to a complete integration with well-established Si based micro-electronics technology.

The limitation of thin film based sensor was described. A number of thin film sensors might be recognized from sensor arrays which yield slightly different responses to various target gases. The availability of thin film gas sensor potentially creates a complicated selection problem, and is more important in view of cost and technology limitations. Many researchers have self-sufficiently confirmed practical limitations to thin film gas detection at low temperature and have attributed it to the requirement of high activation energy which can be attained only at elevated temperatures. A reduction in the number of sensors to be involved in E-Nose is advantageous due to several reasons as discussed. Sensors which exhibit an insignificant response to target gases, increase variance (noise) in E-Nose and do not assist pattern recognition process. Furthermore, sensors exhibiting similar responses to the target gases provide no additional information and are redundant.

In future, low temperature operation of the thin film sensors is an attractive proposition for the industry since it not only holds a promise to cut down the costs but also overcome technological limitations of miniature heaters of high wattage. In order to identify the target gases other classification technique such as artificial neural network approach is required where the selected features/variables obtained from principal component analysis (PCA) could be used as input features, and will be carried out in future. Therefore, a new methodology or novel design approach is essentially required in order to fulfil the essential requirements of future sensor in the market.

5. Metal sulphide, metal telluride and metal selenide thin film based solar cells

Metal chalcogenide materials are considered as excellent absorber materials in photovoltaic cell applications. These materials exhibited excellent absorption co-efficient and suitable band gap value to adsorb the maximum number of photons from sun radiation. Photovoltaic cell can be used to convert sunlight into electricity. These materials have a several advantages such as flexible, lower in weight, have less drag and very thin layer (from nanometer to micrometers). Preparation of the films has been reported by many researchers via different deposition methods. The properties of obtained films were studied by using various tools. The obtained experimental findings revealed that these materials could be classified into two groups, namely p-type and n-type materials. Experimental results confirmed that electron (n-type material) can absorb the energy from photons, following that, jump to the p-type materials, to produce electric potential.

Metal chalcogenide materials are considered as excellent absorber materials in photovoltaic cell applications [105, 106]. These materials exhibited excellent absorption co-efficient and suitable band gap value to adsorb the maximum number of photons from sun radiation [107, 108]. Photovoltaic cell can be used to convert sunlight into electricity. These materials have a several advantages such as flexible, lower in weight, have less drag and very thin layer (from nanometer to micrometers). Preparation of the films has been reported by many researchers via different deposition methods [109]. The properties of obtained films were studied by using various tools [110]. The obtained experimental findings revealed that these materials could be classified into two groups [111, 112], namely p-type and n-type materials. Experimental results confirmed that electron (n-type material) can absorb the energy from photons, following that, jump to the p-type materials, to produce electric potential.

Based on the global photovoltaic market [113], the market shares of silicon based solar cell decreased from 92% (in 2014) to 73.3% in 2020. Silicon based solar cell accountable for the highest percentage of market share due to the abundant raw material availability and high efficiency value. The thin film based solar cells increased from 2014 (7%) to 2020 (10.4%). Solar cell market is expected to growth rapidly due to the rising demand for commercial, residential and utility applications. According to the market share of thin film technologies [114], there are three common thin film materials such as amorphous silicon, cadmium telluride and copper indium gallium selenide. Amorphous silicon based solar cell was the oldest thin film technologies, and it dominates overall market from 2000 to 2003. This type of solar cell can absorb a wide range of the light spectrum, did excellent in low light, but loses efficiency rapidly. The CdTe films have been deposited successfully onto glass. Quaternary thin films such as copper indium gallium selenide were prepared via co-evaporation method. The global demand for CdTe films and CIGS films was expected to drive the market start from 2004 and onwards [114].

The cadmium telluride thin films could be used as solar absorber due to suitable band gap value and high absorption coefficient in the visible light region [115]. The materials have high absorption coefficient was able a low absorber thickness (about 1 μm) to absorb sufficient sunlight. Generally, several researchers reported the synthesis of CdTe films by using various deposition methods such as chemical bath deposition [116], spray pyrolysis [117], thermal evaporation [118], molecular beam epitaxy [119], close space sublimation [120], pulsed laser deposition method [121], hydrothermal method, electrochemical deposition technique. Researchers pointed out that the CdTe films deposited onto glass substrates showed some problems such as heavy and fragile. Currently, more and more research activities are focusing on the synthesis of CdTe films onto metal foils in order to lower the investment in equipment and infrastructure. The thin film deposited onto flexible substrates could be folded in any shape, and the researcher concluded that the supporting structure requirements are minimum if compared to heavy glass substrates. **Table 2** showed the advantages, limitations, power conversion efficiency of CdTe films. Also, the solar power plant was described in the table. So far, the First Solar Company is the main producer of CdTe film.

Advantages	Limitations	Solar power plant	Power conversion efficiency (%)
CdTe has band gap about 1.5 eV, it can absorb sunlight at close to ideal wavelength, it captures energy at shorter wavelength.	Toxic effect of cadmium	The Topaz Solar Farm was located in California, United States. The photovoltaic power station includes 9 million CdTe thin film modules [122].	19% as reported by Gloeckler and co-workers [123]
The cadmium is abundant	Very limited availability of tellurium	In the Desert Sunlight Solar Farm (California, United States), it employed 8.8 million CdTe film modules [124].	13.3% as highlighted by Kamala and coworkers [125]
CdTe film based solar cell showed the shortest energy payback time and the smallest carbon footprint.	It does not remain stable under severe stress	The Waldpolenz Solar Park was located in Germany, has used CdTe film modules, was 52megawatt photovoltaic power station [126].	15% as pointed out by Devendra and co-workers [127]
	It is very important to enhance the efficiency of solar cells	Templin solar power plant was located in Germany, has installed more than 1.5 million CdTe film modules [128].	17.8% as concluded by Deb and coworkers [129]
			9.59% as described by Xixing and coworkers [130]

Table 2.
Advantages, limitations, power conversion efficiencies and CdTe film based solar power plant.

The copper indium gallium selenide (CIGS) thin films have been prepared by using different deposition methods such as thermal evaporation method [131], spray pyrolysis [132], solvothermal method [133], physical vapor deposition [134], and electro deposition method [135]. **Table 3** showed the advantages, limitations and power conversion efficiency of CIGS thin films. These films showed p-type absorbing layer materials and the tunable band gap (1.07–1.7 eV) value [141]. Researcher highlighted that there are 99% of the light will be successfully absorbed in the first micrometer of the materials [142]. The solar cell is classified as heterojunction structures [143]. Generally, the junction is produced between thin films having various band gap values. Experimental results showed that the addition of small amount of gallium can improve the voltage, boost band gap value and enhance the power conversion efficiency of solar cell [144]. There are several companies produced CIGS solar cell such as Solar Frontier, Solyndra, SoloPower, Global Solar, SulfurCell, MiaSole and Nanosolar. The solar cell showed open circuit voltage, short circuit current and the maximum power values of 5 V Dc, 95 mA and 0.25 watts, respectively.

The copper rich p-type $CuInS_2$ films were synthesized by using thermal co-evaporation method. The obtained results showed that small (less than 10%) solar to electrical conversion losses when the copper to indium ration between 1 and 1.8. The highest power conversion efficiency was 10.2% as reported by Scheer and co-workers [145].

The chemical bath deposition was used to produce $Ni_3Pb_2S_2$ thin films [146]. The photovoltaic parameters such as open circuit voltage (0.61 V), short circuit current density (9.9 mA/cm^2), fill factor (0.47) and power conversion efficiency (2.7%) were studied. The band gap was calculated based on the absorption spectra and was about 1.4 eV.

The atomic layer deposition was employed to produce SnS films [147] as highlighted by Rafael and co-workers. These materials are non-toxic solar cell, and the power conversion efficiency was 4.36%. Vera and co-workers [148] reported that SnS heterojunction solar cell was made, and reached power conversion efficiency about 3.88%.

Advantages	Limitations	Power conversion efficiency (%)
CIGS thin films have been deposited onto substrates (flexible)	Less efficient if compared to silicon based solar panels	Conventional solar cell: 22.67%. Adding the BSF (PbS) layer in solar cell: 24.22% as reported by Barman and Kalita [136]
The active layer could be deposited in polycrystalline form.	Higher production costs if compared to other thin film technologies.	The highest efficiency is 25.5% as highlighted by Sobayel and coworkers [137]
Much lower level of cadmium will be used during the synthesis of thin films	Complex structures	The highest power conversion efficiency was 26.4% as concluded by Sobayel and co-workers [138]
CIGS thin films based solar panel indicated better resistance to heat.		Boubakeur et al. have achieved power conversion efficiency of 21.08%. [139]
Much less expensive if compared to silicon based solar cells.		Nour and Patane reported the highest power conversion efficacy about 24.5%. [140]

Table 3.
The advantages, limitations and power conversion efficiency of CIGS thin films.

The performance of p-type InSe films for solar cell was reported. The open circuit voltage (0.55 V), short circuit current density (7.09 mA/cm2), fill factor (53.85%, and power conversion efficiency (0.52%) were highlighted. Researchers explained that higher series resistance and reduced shunt resistance lead to lower value of efficiency. The band gap values are in the range of 1.75–1.95 eV in as-deposited films, annealed films at 250 and 300°C as concluded by Teena and co-workers [149].

The electrochemical technique was used to produce CdSe film MnCdSe films as described by Shinde and co-workers [150]. XRD analysis showed the obtained films are polycrystalline with hexagonal crystal phase. The SEM images revealed that nanosphere morphology and nanonest structure for CdSe and MnCdSe films respectively. The band gap value was measured, and the reduced from 1.81 eV (CdSe) to 1.6 eV (MnCdSe). The fill factor and power conversion efficiency of CdSe films 0.71 and 0.67%, respectively. The MnCdSe films showed power conversion efficiency about 0.37%.

thin films	Power conversion efficiency (%)	References
Cu_2ZnSnS_4	5.74	Kazuo and co-workers [155]
Cu_2ZnSnS_4	2.62	Hironori and co-workers [156]
Cu_2ZnSnS_4	6.8	Wang and co-workers [157]
Cu_2ZnSnS_4	4.1	Schubert and co-workers [158]
Cu_2ZnSnS_4	0.23	Chet and co-workers [159]
Cu_2ZnSnS_4	3.2	Jonathan and co-workers [160]
Cu_2ZnSnS_4	3.4	Ennaoui and co-workers [161]
Cu_2ZnSnS_4	0.396	Sawanta and co-workers [162]
Cu_2ZnSnS_4	6.03	Tsukasa and co-workers [163]
Cu_2ZnSnS_4	0.12	Shinde and co-workers [164]
CuS	0.39	Donghyeok and co-workers [165]
CdS	8	Karl [166]
ZnS	8.83	Qiu and co-workers [167]
PbS	2.02	Omer and co-workers [168]
PbS:Mo	2.16	Omer and co-workers [168]
Sb2Se3	7.6	Wen and co-workers [169]
Sb_2Se_3	5.93	Liang and co-workers [170]
Sb_2Se_3	5.6	Chao and co-workers [171]
$CuInTe_2$	3.8	Manorama and co-workers [172]
$CuInTe_2$	4.13	Lakhe and co-workers [173]
$CuInTe_2$	1.22	Jia and co-workers [174]
$CuInSe_2$	1.75	Hyun and co-workers [175]
$CuInSe_2$	2	Se and co-workers [176]
$CuInSe_2$	4.57	Prabukanthan and co-workers [177]
$MnCuInSe_2$	6.38	Prabukanthan and co-workers [177]

Table 4.
Power conversion efficiencies of different types of thin films.

The ternary compound such as Cu_2SnS_3 (CTS) films showed high absorption coefficient (10^4 cm^{-1}) and wider range of band gap energy (0.9–1.7 eV). Researchers reported that easy to control the secondary phase during the synthesis of CTS films. The formation of cubic, monoclinic, tetragonal and orthorhombic structure strongly depended on deposition method and annealing process. The magnetron sputtering method was used to produce CTS films. The films reached the highest power conversion efficiency about 2.2%, due to the formation of pure phase of CTS, lowest sheet resistance (8.2 Ω/cm^2), highest shunt resistance (111.1 Ω/cm^2) and uniform morphology [151]. The p-type CTS films have been produced via co-evaporation method [152]. The photovoltaic parameters such as open circuit voltage (248 mV), short circuit current density (33.5 mA/cm^2), fill factor (0.439) and power conversion efficiency (3.66%) were highlighted. Mingrui and co-workers [153] described the preparation of CTS films by using sputtering method. The films prepared at 2812 seconds indicated the highest efficiency value (2.39%), with fill factor (39.7%), open circuit current voltage (208 mV) and short circuit current density (28.92 mA/cm^2).

The Cu_4SnS_4 films showed p-type electrical conductivity and the band gap values (0.93–1.84 eV). Chen et al., have reported the synthesis of thin films by a combination of mechanochemical and doctor blade techniques [154]. The highest power conversion efficiency reached 2.34%. The influence of the film thickness on the properties of samples was study. Based on the absorption spectra, the absorption edge moved towards longer wavelength with increasing the film thickness (0.25–1 μm). Also, band gap reduced (1.47–1.21 eV) due to reduction of structural disorder and the increase in the crystalline size.

The **Table 4** showed the power conversion efficiency of the various thin films. The obtained experimental results confirmed that metal sulfide, metal selenide and metal telluride thin films could be used in solar cell applications. The photovoltaic parameters were strongly depended on various experimental conditions. Researchers also highlighted a lot of research activities have been carried put in order to enhance the power conversion efficiency of thin film based solar cell.

6. Conclusions

Chalcogenide thin films have received a great deal of attention for decades due to some unique properties. The thin film based supercapacitor can have store the electrical energy for all the electronic devices to stabilize the power supply. Metal nitrates have gotten a lot of interest as supercapacitors electrodes due to showed higher electrical conductivity, higher specific capacitance, good energy density, and excellent mechanical stability. Perovskite solar cell indicated higher power conversion efficiency value. The organic inorganic hybrid perovskite solar cells are very simple, and prepared by sandwiching a perovskite absorber layer between the electron transport layer and hole transport layer, reached power conversion efficiency exceeding 20%. The thin film-based sensors showed high sensitivity and low manufacturing cost. In future, low temperature operation of the thin film sensors is an attractive proposition for the industry. The market shares of silicon based solar cell decreased, while thin film based solar cells increased in the global photovoltaic market due to the low material consumption, low manufacturing cost, shorter energy pack back period. Solar cell market is expected to growth rapidly due to the rising demand for commercial, residential and utility applications.

Acknowledgements

The author (HO SM) gratefully acknowledge the financial support provided by the INTI International University.

Conflict of interest

The authors declare no conflict of interest.

Author details

Ho Soonmin[1*], Immanuel Paulraj[2], Mohanraj Kumar[3], Rakesh K. Sonker[4] and Pronoy Nandi[5]

1 Faculty of Health and Life Science, INTI International University, Putra Nilai, Negeri Sembilan, Malaysia

2 Department of Physics, National Changhua University of Education, Changhua, Taiwan

3 Department of Environmental Engineering and Management, Chaoyang University of Technology, Taichung, Taiwan

4 Department of Physics, Acharya Narendra Dev College, University of Delhi, Delhi, India

5 Institute of Physics, Sachivalaya Marg, Bhubaneswar, Odisha, India

*Address all correspondence to: soonmin.ho@newinti.edu.my

References

[1] Xing C, Lei Y, Liu M, Wu S, He W. Environment-friendly Cu-based thin film solar cells: materials, devices and charge carrier dynamics. Physical Chemistry Chemical Physics. 2021;**23**:16469-16487

[2] Kassim A, Saravanan N, Ho SM, Lim K. SEM, EDAX and UV-visible studies on the properties of Cu_2S thin films. Chalcogenide Letters. 2011;**8**:405-410

[3] Ngai CF, Anuar K, Ho SM, Tan W. Influence of triethanolamine on the chemical bath deposited NiS thin films. American Journal of Applied Sciences. 2011;**8**:359-361

[4] Faiz R. Zinc oxide light emitting diodes: A review. Optical Engineering. 2019;**58**. DOI: 10.1117/1.OE.58.1.010901

[5] Ho SM. Influence of complexing agent on the growth of chemically deposited $Ni_3Pb_2S_2$ thin films. Oriental Journal of Chemistry. 2014;**30**:1009-1012

[6] Atan S, Ho SM, Anuar K, Saravanan N. Effect of deposition period and pH on chemical bath deposited Cu_4SnS_4 thin films. Philippine Journal of Science. 2009;**138**:161-168

[7] Jesu A, Najla A, Khan A, Mohd S, Algarni H. High sensitive samarium-doped ZnS thin films for photo-detector applications. Optical Materials. 2021;**122**. DOI: 10.1016/j.optmat.2021.111649

[8] Tan WT, Ho SM, Anuar K. Composition, morphology and optical characterization of chemical bath deposited ZnSe thin films. European Journal of Applied Sciences. 2011;**3**:75-80

[9] Shinde K, Dhaygude D, Fulari V, Chikode P. Structural, morphological, optical and hologram recording of the CdS and ZnS thin films by double exposure digital holographic interferometry technique. Journal of Materials Science: Materials in Electronics. 2017;**28**:7385-7392

[10] Ramos M, Lopez J, Rascon J, Reyes D, Flores P. Structural and optical modifications of CdS properties in CdS-Au thin films prepared by CBD. Results in Physics. 2021;**22**. DOI: 10.1016/j.rinp.2021.103914

[11] Teo D, Anuar K, Saravanan N, Ho SM, Tan W. Chemical bath deposition of nickel sulphide (Ni_4S_3) thin films. Leonardo Journal of Sciences. 2010;**16**:1-12

[12] Jelas M, Anuar K, Ho SM, Tan W, Gwee SY. Effects of Deposition Period on the Properties of FeS_2 Thin Films by Chemical Bath Deposition Method. Thammasat International Journal of Science and Technology. 2010;**15**:62-69

[13] Anuar K, Ho SM, Saravanan N, Noraini K. XRD and AFM studies of ZnS thin films produced by electro deposition method. Arabian Journal of Chemistry. 2010;**3**:243-249

[14] Atef S, Sayed M. Electrodeposition, characterization and photo electrochemical properties of CdSe and CdTe. Ain Shams Engineering Journal. 2014;**6**:341-346

[15] Tuyen T, Sudam C, Ivet K, Oscar M, Ramon T. Electrodeposition of antimony selenide thin films and application in semiconductor sensitized solar cells. ACS Applied Materials & Interfaces. 2014;**6**:2836-2841

[16] Sreeja V, Sabitha P, Anila E, Radhakrishnan P. Nonlinear optical

characterization of ZnS thin film synthesized by chemical spray pyrolysis method. AIP Conference Proceedings. 2015;**1620**. DOI: 10.1063/1.4898292

[17] Mahalingam T, Lee S, Kim Y, Moon S, Kathalingam A. Studies of electro synthesized zinc selenide thin films. Journal of New Materials for Electrochemical Systems. 2007;**10**:15-19

[18] Fida M, Tahir M, Zeb M, Ali S, Sabri M, Sarker R. Synergistic enhancement in the microelectronic properties of poly-(dioctylfluorene) based Schottky devices by CdSe quantum dots. Scientific Reports. 2020;**10**. DOI: 10.1038/s41598-020-61602-1

[19] Seok Y, Seo W, Lee S, Shim W. Preparation of $CuInSe_2$ thin films through metal organic chemical vapor deposition method by using di-μ-methylselenobis(dimethylindium) and bis(ethylisobutyrylacetato) copper(II) precursors. Thin Solid Films. 2006;**515**:1544-1547

[20] Marianna K, Mikko R, Markku L. Thin Film Deposition Methods for CuInSe $_2$ Solar Cells. Critical Reviews in Solid State and Materials Sciences. 2007;**30**. DOI: 10.1080/10408430590918341

[21] Azhar K, Sabah H, Khalaf K. The effects of sputtering time on Cds thin film solar cell deposited by DC plasma sputtering method. Engineering and Technology Journal. 2018;**36**. DOI: 10.30684/etj.36.2C.5

[22] Oday A, Jafar R, Adnan M, Nadir F, Chiad S. Synthesize of CdS and CdS: V films via spray pyrolysis technique: Morphology and optical properties. Design Engineering. 2020;**5**:356-365

[23] Atan S, Ho SM, Kassim A, Nagalingam S. X-ray diffraction and atomic force microscopy studies of chemical bath deposited FeS thin films. Studia Universitatis Babes-Bolyai Chemia. 2010;**55**:5-11

[24] Ho SM, Anuar K. Deposition and characterization of MnS thin films by chemical bath deposition method. International Journal of Chemistry Research. 2010;**1**:1-5

[25] Sharif S, Khuram S, Saeed A, Rana F, Azad M. In situ synthesis and deposition of un-doped and doped magnesium sulfide thin films by green technique. Optik. 2019;**182**:739-744

[26] Okoli D. Optical properties of chemical bath deposited magnesium sulphide thin films. Chemistry and Materials Research. 2015;7:61-67

[27] Rasaq A, Kola R, Adeniyi T, Talabi T. Synthesis and characterization of ZnSe thin films deposited by thermal vacuum evaporation method for photovoltaic application. Applied Journal of Environmental Engineering Science. 2020;**6**:227-237

[28] Anuar K, Ho SM, Noraini K, Saravanan N. XRD and AFM studies of ZnS thin films produced by electrodeposition method. Arabian Journal of Chemistry. 2010;**3**:243-249

[29] Ghezali K, Mentar L, Azizi A, Boudine B. Electrochemical deposition of ZnS thin films and their structural, morphological and optical properties. Journal of Electroanalytical Chemistry. 2017;**794**:212-220

[30] Patil S, Fulari J, Lohar M. Structural, morphological, optical and photoelectrochemical cell properties of copper oxide using modified SILAR method. Journal of Materials Science: Materials in Electronics. 2016;**27**:9550-9557

[31] Lee W, Wang X. Structural, Optical, and Electrical Properties of Copper Oxide Films Grown by the SILAR Method with Post-Annealing. Coatings. 2021;**11**. DOI: 10.3390/coatings11070864

[32] Ho SM. Properties study of SILAR deposited cobalt selenide thin films. International Journal of Research and Review. 2021;**8**. DOI: 10.52403/ijrr.20211216

[33] Chavan T, Yadav A, Kamble S, Sabah A, Insik I, Cho E, et al. Electrochemical supercapacitive studies of chemically deposited $Co_{1-x}Ni_xS$ thin films. Materials Science in Semiconductor Processing. 2020;**107**. DOI: 10.1016/j.mssp.2019.104799

[34] Zhao J, Tian Y, Liu A, Zhao Z, Song L. The NiO electrode materials in electrochemical capacitor: A review. Materials Science in Semiconductor Processing. 2019;**96**:78-90

[35] Wiston B, Ashok M. Microwave-assisted synthesis of cobalt-manganese oxide for supercapacitor electrodes. Materials Science in Semiconductor Processing. 2019;**103**. DOI: 10.1016/j.mssp.2019.104607

[36] Sarkar S, Howli P, Das B, Das N, Samanta M. Novel quaternary chalcogenide/reduced graphene oxide-based asymmetric supercapacitor with high energy density. ACS Applied Materials Interfaces. 2019;**9**:22652-22664

[37] Velanganni S, Pravinraj S, Immanuel P. Nanostructure CdS/ZnO heterojunction configuration for photocatalytic degradation of Methylene blue. Physica B: Condensed Matter. 2018;**534**:56-62

[38] Zahirullah S, Immanuel P, Pravinraj S, Inbaraj H, Joseph J. Synthesis and characterization of Bi doped ZnO thin films using SILAR method for ethanol sensor. Materials Letters. 2018;**230**:1-4

[39] Immanuel P, Prakash A, Raja M. Ethanol sensing of V_2O_5 thin film prepared by spray pyrolysis technique: Effect of substrate to nozzle distance. AIP Conference Proceedings. 2017;**1832**. DOI: 10.1063/1.4980482

[40] Immanuel P, Mohan R. Effect of process temperature on the preparation of V_2O_5 thin films by spray pyrolysis method for ethanol sensing application. Materials Focus. 2016;**5**:362-367

[41] Dai C, Chien P, Lin Y, Chou W, Li P, Lin T. Hierarchically structured Ni_3S_2/carbon nanotube composites as high performance cathode materials for asymmetric supercapacitors. ACS Applied Materials Interfaces. 2013;**5**:12168-12174

[42] Xiao Y, Fang S, Su D, Wang X, Zhou L, Wu S, et al. In suit growth of ultra-dispersed $NiCo_2S_4$ nanoparticles on graphene for asymmetric supercapacitors. Electrochimica Acta. 2015;**176**:44-50

[43] Mukkabla R, Deepa M, Kumar S. Poly(3,4-ethylenedioxypyrrole) Enwrapped Bi_2S_3 Nanoflowers for Rigid and Flexible Supercapacitors. Electrochimica Acta. 2015;**164**:171-181

[44] Balogun M, Zeng Y, Luo Y, Qiu W, Titus K, Tong Y. Three-dimensional nickel nitride (Ni_3N) nanosheets: free standing and flexible electrodes for lithium ion batteries and supercapacitors. Journal of Materials Chemistry A. 2016;**4**:9844-9849

[45] Vivek E, Arulraj A, Khalid M, Vetha P. Facile synthesis of 2D $Ni(OH)_2$ anchored g-C_3N_4 as electrode material for high-performance supercapacitor.

Inorganic Chemistry Communications. 2012;**130**. DOI: 10.1016/j. inoche.2021.108704a

[46] Arulraj A, Mehana U, Ramesh M, Raj A. Metal chalcogenides as counter electrode materials. In: Alagarsamy P, Mohan R, Kandasamy J, editors. Counter Electrode for Dye-Sensitized Solar Cells. 1st ed. New York: Jenny Stanford Publishing; 2021. pp. 126-156. DOI: 10.1201/9781003110774

[47] Krishnan G, Arulraj A, Priyanka J, Khalid M. 2D materials for supercapacitor and supercapattery applications. In: Singh L, Durga M, editors. Adapting 2D Nanomaterials for Advanced Application. Washington DC: ACS Publication; 2020. pp. 33-47. DOI: 10.1021/bk-2020-1353

[48] Krishnan S, Arulraj A, Khalid M, Reddy M. Energy storage in metal cobaltite electrodes: Opportunities & challenges in magnesium cobalt oxide. Renewable and Sustainable Energy Reviews. 2021;**141**. DOI: 10.1016/j. rser.2021.110798

[49] Guo Y, Hong F, Wang Y, Li Q, Dai R, Meng J, et al. Multicomponent hierarchical Cu-doped NiCo-LDH/ CuO double arrays for ultralong-life hybrid fiber supercapacitor. Advanced Functional Materials. 2019;**29**. DOI: 10.1002/adfm.201809004

[50] Zhang Y, Hu Y, Wang Z, Zhu X, Luo B, Han H, et al. Lithiation-Induced Vacancy Engineering of Co_3O_4 with Improved Faradic Reactivity for High-Performance Supercapacitor. Advanced Functional Materials. 2020;**30**. DOI: 10.1002/adfm.202004172

[51] Bandgar B, Vadiyar M, Ling Y, Chang J, Han S, Anil V. Metal precursor dependent synthesis of $NiFe_2O_4$ thin films for high-performance flexible symmetric supercapacitor. ACS Applied Energy Materials. 2018;**1**:638-648

[52] Immanuel P, Chang H, Mohanraj K, Kumar S. Effect of Cr doping on Mn_3O_4 thin films for high-performance supercapacitors. Journal of Materials Science: Materials in Electronics. 2021;**32**:3732-3742

[53] Jesuraj S, Haris M, Immanuel P. Structural and optical properties of pure Nio and Li-doped nickel oxide thin films by sol-gel spin coating method. International Journal of Science and Research. 2013;**8**:85-87

[54] Kim Y, Shin K. Dopamine-assisted chemical vapour deposition of polypyrrole on graphene for flexible supercapacitor. Applied Surface Science. 2021;**547**. DOI: 10.1016/j. apsusc.2021.149141

[55] Yu S, Xiong X, Jun M, Qian H. One-step preparation of cobalt nickel oxide hydroxide @cobalt sulfide heterostructure film on Ni foam through hydrothermal electrodeposition for supercapacitors. Surface and Coatings Technology. 2021;**426**. DOI: 10.1016/j. surfcoat.2021.127791

[56] Immanuel P, Mohanraj K, Senguttuvan G. Enhanced activity of chemically synthesized nanorod Mn3O4 thin films for high performance supercapacitors. International Journal of Thin Film Science and Technology. 2020;**9**:57-67

[57] Vivek E, Arulraj A, Syam G, Vetha P, Khalid M. Novel nanostructured $Nd(OH)_3$/ $g\text{-}C_3N_4$ nanocomposites (nanorolls anchored on nanosheets) as reliable electrode material for supercapacitors. Energy & Fuels. 2021;**35**:15205-15212

[58] Arulraj A, Ramesh M, Rajeshkumar V, Ilayaraja N. Direct

synthesis of cubic shaped Ag_2S on Ni mesh as binder-free electrodes for energy storage applications. Scientific Reports. 2019;**9**. DOI: 10.1038/s41598-019-46583-0

[59] Moller C. Crystal structure and photoconductivity of cæsium plumbohalides. Nature. 1958;**182**. DOI: 10.1038/1821436a0

[60] Kim Y, Lee J, Jung S, Shin H, Park N. High-efficiency perovskite solar cells. Chemical Reviews. 2020;**120**:7867-7918

[61] Nandi P, Giri C, Manju U, Rath S, Topwal D. $CH_3NH_3PbI_3$, a potential solar cell candidate: Structural and spectroscopic investigations. The Journal of Physical Chemistry A. 2016;**120**:9732-9739

[62] Nandi P, Mahana S, Welter E, Topwal D. Probing the role of local structure in driving the stability of halide perovskites $CH_3NH_3PbX_3$. The Journal of Physical Chemistry C. 2021;**125**:24655-24662

[63] Nandi P, Pandey K, Giri C, Vijay S, Manju U, Mahanti D, et al. Probing the electronic structure of hybrid perovskites in the orientationally disordered cubic phase. The Journal of Physical Chemistry Letters. 2020;**11**:5719-5727

[64] Menendez E, Palacios P, Wahnon P, Conesa C. Self-consistent relativistic band structure of the CH3NH3PbI3 perovskite. Physical Review B. 2014;**90**. DOI: 10.1103/PhysRevB.90.045207

[65] Xiao Z, Yan Y. Progress in theoretical study of metal halide perovskite solar cell materials. Advanced Energy Materials. 2017;7. DOI: 10.1002/aenm.201701136

[66] Motta C, Fedwa E, Kais S, Tabet N, Fahhad A, Sanvito S. Revealing the role of organic cations in hybrid halide perovskite $CH_3NH_3PbI_3$. Nature Communications. 2015;**6**. DOI: 10.1038/ncomms8026

[67] Li Y, Ji L, Liu R, Zhang C, Mak C, Zou X, et al. A review on morphology engineering for highly efficient and stable hybrid perovskite solar cells. Journal of Materials Chemistry A. 2018;**6**:12842-12875

[68] Nandi P, Giri C, Topwal D. Understanding the origin of broad-band emission in $CH_3NH_3PbBr_3$. Journal of Materials Chemistry C. 2021;**9**:2793-2800

[69] Nandi P, Li Z, Kim Y, Ahn T, Park N, Shin H. Stabilizing mixed halide lead perovskites against photoinduced phase segregation by A-site cation alloying. ACS Energy Letters. 2021;**6**:837-847

[70] Ogomi Y, Kenji Y, Pandey S, Shuzi H, Ma T, Syota T, et al. $CH_3NH_3Sn_xPb_{(1-x)}I_3$ perovskite solar cells covering up to 1060 nm. The Journal of physical Chemistry Letters. 2014;**5**:1004-1011

[71] Sarkar S, Priya M. Role of the A-site cation in determining the properties of the hybrid perovskite CH3NH3Pb. Physical Review B. 2017;**95**. DOI: 10.1103/PhysRevB.95.214118

[72] Xing G, Mathews N, Sun S, Lim S, Lam M, Sum T, et al. Long-range balanced electron- and hole-transport lengths in organic-inorganic $CH_3NH_3PbI_3$. Science. 2013;**342**:344-347

[73] Zhumekenov A, Makhsud I, Azinul M, Erkki A, Omar F, Bakr O. Formamidinium lead halide perovskite crystals with unprecedented long carrier dynamics and diffusion length. ACS Energy Letters. 2016;**1**:32-37

[74] Wehrenfennig C, Liu M, Henry J, Johnson M, Laura M. Charge carrier recombination channels in the

low-temperature phase of organic-inorganic lead halide perovskite thin films. AIP APL Materials. 2014;**2**. DOI: 10.1063/1.4891595

[75] Govinda S, Kore P, Swain D, Akmal H, De C, Tayur N, et al. Critical comparison of $FAPbX_3$ and $MAPbX_3$ (X = Br and Cl): How do they differ. The Journal of Physical Chemistry C. 2018;**122**:13758-13766

[76] Nandi P, Dinesh T, Park N, Shin H. Organic-inorganic hybrid lead halides as absorbers in perovskite solar cells: a debate on ferroelectricity. Journal of Physics D: Applied Physics. 2020;**53**. DOI: 10.1088/1361-6463/abb047

[77] Waikar M, Raste P, Sonker K, Gupta V, Tomar M, Shirsat D. E nhancement in NH_3 sensing performance of ZnO thin-film via gamma-irradiation. Journal of Alloys and Compounds. 2020;**830**. DOI: 10.1016/j.jallcom.2020.154641

[78] Liu H, Yu H, Wang J, Xia F, Wang C, Xiao J. $LaNbO_4$ as an electrode material for mixed-potential CO gas sensors. Sensors and Actuators B: Chemical. 2022;**352**. DOI: 10.1016/j.snb.2021.130981

[79] Sonker R, Yadav B, Sabhajeet S. TiO_2–PANI nanocomposite thin film prepared by spin coating technique working as room temperature CO_2 gas sensing. Journal of Materials Science: Materials in Electronics. 2016;**27**:11726-11732

[80] Sonker K, Yadav C, Gupta V, Tomar M, Sharma A. Experimental investigations on NO_2 sensing of pure ZnO and PANI–ZnO composite thin films. RSC Advances. 2016;**6**:56149-56158

[81] Tyagi P, Tomar M, Vinay G, Anjali S. A comparative study of RGO-SnO_2 and MWCNT-SnO_2 nanocomposites based SO_2 gas sensors. Sensors and Actuators B: Chemical. 2017;**248**:980-986

[82] Mousavi H, Yadollah M, Ali A, Saberi H. Enourmous enhancement of Pt/SnO2 sensors response and selectivity by their reduction to CO in automotive exhaust gas pollutants including CO, NOx and C3H8. Applied Surface Science. 2021;**546**. DOI: 10.1016/j.apsusc.2021.149120

[83] Sabhajeet S, Yadav C, Sonker R. Sol-gel formed spherical nanostructured titania based liquefied petroleum gas sensor. AIP Conference Proceedings. 2018;**1953**. DOI: 10.1063/1.5032413

[84] Chowdhuri A, Kumar S, Vinay G, Sreenivas K. Contribution of adsorbed oxygen and interfacial space charge for enhanced response of SnO_2 sensors having CuO catalyst for H_2S gas. Sensors and Actuators B: Chemical. 2010;**145**:155-166

[85] Sonker RK, Yadav BC. Chemical route deposited SnO_2, SnO_2-Pt and SnO_2-Pd thin films for LPG detection. Advanced Science Letters. 2014;**20**:1023-1027

[86] Patil PT, Anwane RS, Kondawar SB. Development of electrospun polyaniline/ZnO composite nanofibers for LPG sensing. Procedia Materials Science. 2015;**10**:195-204

[87] Dhawale DS, Salunkhe R, Patil UM, Gurav KV, More AM, Lokhande CD. Room temperature liquefied petroleum gas (LPG) sensor based on p-polyaniline/n-TiO2 heterojunction. Sensors and Actuators B: Chemical. 2008;**134**:988-992

[88] Patil LA, Suryawanshi DN, Pathan IG, Patil DM. 2013. Nickel doped spray pyrolyzed nanostructured TiO2 thin films for LPG gas sensing. Sensors and Actuators B: Chemical. 2013;**176**:514-521

[89] Ladhe RD, Gurav KV, Pawar SM, Kim JH, Sankapal BR. p-PEDOT: PSS as

a heterojunction partner with n-ZnO for detection of LPG at room temperature. Journal of Alloys and Compounds. 2012;**515**:80-85

[90] Sonker RK, Singh M, Kumar U, Yadav BC. MWCNT doped ZnO nanocomposite thin film as LPG sensing. Journal of Inorganic and Organometallic Polymers and Materials. 2016;**26**:1434-1440

[91] Gautam C, Tiwary CS, Machado LD, Jose S, Ozden S, Biradar S, et al. Synthesis and porous h-BN 3D architectures for effective humidity and gas sensors. RSC Advances. 2016;**6**:87888-87896

[92] Sonker RK, Yadav BC, Dzhardimalieva GI. Preparation and properties of nanostructured PANI thin film and its application as low temperature NO2 sensor. Journal of Inorganic and Organometallic Polymers and Materials. 2016;**26**:1428-1433

[93] Sonker R, Yadav B, Gupta V, Tomar M. Fabrication and characterization of ZnO-TiO2-PANI (ZTP) micro/nanoballs for the detection of flammable and toxic gases. Journal of Hazardous Materials. 2019;**370**:126-137

[94] Sonker R, Yadav B, Gupta V, Tomar M. Synthesis of CdS nanoparticle by sol-gel method as low temperature NO2 sensor. Materials Chemistry and Physics. 2020;**239**. DOI: 10.1016/j.matchemphys.2019.121975

[95] Herrán J, Mandayo GG, Ayerdi I, Castano E. Influence of silver as an additive on BaTiO3–CuO thin film for CO2 monitoring. Sensors and Actuators B: Chemical. 2008;**129**:386-390

[96] Chapelle A, Oudrhiri-Hassani F, Presmanes L, Barnabé A, Tailhades P. CO2 sensing properties of semiconducting copper oxide and spinel ferrite nanocomposite thin film. Applied Surface Science. 2010;**256**:4715-4719

[97] Krishnakumar T, Jayaprakash R, Prakash T, Sathyaraj D, Donato N. CdO-based nanostructures as novel CO2 gas sensors. Nanotechnology. 2011;**22**. DOI: 10.1088/0957-4484/22/32/325501

[98] Chiang CJ, Tsai K, Lee Y, Lin H, Yang Y, Shih C, et al. In situ fabrication of conducting polymer composite film as a chemical resistive CO2 gas sensor. Microelectronic Engineering. 2013;**111**:409-415

[99] Fan K, Qin H, Wang L, Ju L, Hu J. CO2 gas sensors based on La1– xSrxFeO3 nanocrystalline powders. Sensors and Actuators B: Chemical. 2013;**177**:265-269

[100] Habib M, Hussain SS, Riaz S, Naseem S. Preparation and characterization of ZnO nanowires and their applications in CO2 gas sensors. Materials Today: Proceedings. 2015;**2**:5714-5719

[101] Jeong YJ, Balamurugan C, Lee DW. Enhanced CO2 gas-sensing performance of ZnO nanopowder by La loaded during simple hydrothermal method. Sensors and Actuators B: Chemical. 2016;**229**:288-296

[102] Sonker R, Sabhajeet S, Yadav B, Johari R. Liquefied petroleum gas detection using SnO2, PANI-SnO2 and Ag-SnO2 composite film fabricated by chemical route. International Journal of Electroactive Materials. 2017;**5**:6-12

[103] Sabhajeet S, Sonker R, Yadav B. Zn-Doped TiO2 nanoparticles employed as room temperature liquefied petroleum gas sensor. Advanced Science, Engineering and Medicine. 2018;**10**:736-740

[104] Sonker R, Yadav B. Development of Fe2O3–PANI nanocomposite thin film

based sensor for NO2 detection. Journal of the Taiwan Institute of Chemical Engineers. 2017;77:276-281

[105] Choi Y, Park H, Lee N, Kim B, Lee J, Lee G, et al. Deposition of the tin sulfide thin films using ALD and a vacuum annealing process for tuning the phase transition. Journal of Alloys and Compounds. 2022;**896**. DOI: 10.1016/j.jallcom.2021.162806

[106] Alireza G, Sanaz M, Mahdi A, Narges S. Study of optical properties of ZnS and MnZnS (ZnS/MnS) nanostructure thin films; Prepared by microwave-assisted chemical bath deposition method. Materials Chemistry and Physics. 2022;**275**. DOI: 10.1016/j.matchemphys.2021.125103

[107] Chate P, Lakde S, Sathe D. Structural, optical and thermoelectric studies on chemically synthesized annealed antimony sulphide thin films. Optik. 2022;**250**. DOI: 10.1016/j.ijleo.2021.168296

[108] Deepika G, Vishnu C, Sonica U, Singh F, Kumar S, Aman M, et al. Defects engineering and enhancement in optical and structural properties of 2D-MoS_2 thin films by high energy ion beam irradiation. Materials Chemistry and Physics. 2022;**276**. DOI: 10.1016/j.matchemphys.2021.125422

[109] Jrad A, Manel N, Ammar S, Najoua T. Chemical composition, structural, morphological, optical and luminescence properties of chemical bath deposited Fe:ZnS thin films. Optical Materials. 2022;**123**. DOI: 10.1016/j.optmat.2021.111851

[110] Jako S, Kaia T, Oja A, Malle K. Sb_2S_3 thin films by ultrasonic spray pyrolysis of antimony ethyl xanthate. Materials Science in Semiconductor Processing. 2022;**137**. DOI: 10.1016/j.mssp.2021.106209

[111] Rohini M, Oscar G, Ana R, Nair K, Santana E. Thin films of p-SnS and n-Sn_2S_3 for solar cells produced by thermal processing of chemically deposited SnS. Journal of Alloys and Compounds. 2022;**892**. DOI: 10.1016/j.jallcom.2021.162036

[112] Kuang N, Zuo Z, Liu R, Zhao Z. Optimized thermoelectric properties and geometry parameters of annular thin-film thermoelectric generators using n-type $Bi_2Te_{2.7}Se_{0.3}$ and p-type $Bi_{0.5}Sb_{1.5}Te_3$ thin films for energy harvesting. Sensors and Actuators A: Physical. 2021;**332**. DOI: 10.1016/j.sna.2021.113030

[113] Shahariar M, Kazi S, Tanjia C, Kuaanan T, Sieh K. An overview of solar photovoltaic panels' end-of-life material recycling. Energy Strategy Reviews. 2020;**27**. DOI: 10.1016/j.esr.2019.100431

[114] Kelly P. What is thin-film solar? 30 July 2018. Available from: https://www.solarpowerworldonline.com/2018/07/thin-film-solar-stuck-in-second-place-even-with-c-si-tariffs/ [Accessed: December 15, 2021]

[115] Jun W, Mu Y, Qian L, Yang H, Liu T, Fu W. Fabrication of CdTe thin films grown by the two-step electrodeposition technique on Ni foils. Journal of Alloys and Compounds. 2015;**636**:97-101

[116] Deivanayaki S, Ponnuswamy V, Mariappan R, Jayamurugan P. Optical and structural characterization of CdTe thin films by chemical bath deposition technique. Chalcogenide Letters. 2010;7:159-163

[117] Nikale M, Shinde S, Bhosale H, Rajpure K. Physical properties of spray deposited CdTe thin films: PEC performance. Journal of Semiconductors. 2011;**32**. DOI: 10.1088/1674-4926/32/3/033001

[118] Tursun A, Joel N, Zheng X, Helio M, John M, Steven W, et al. Thin-film solar cells with 19% efficiency by thermal evaporation of CdSe and CdTe. ACS Energy Letters. 2020;**5**:892-896

[119] Mikhaylov V, Polyak E. Mass-spectrometry investigation of the kinetics of the molecular-beam epitaxy of CdTe. Journal of Surface Investigation: X-Ray, Synchrotron and Neutron Techniques. 2021;**15**:683-695

[120] Velu R, Shankar B, Nair S, Shanmugam M. Effects of gas-phase and wet-chemical surface treatments on substrates induced vertical, valley–hill & micro-granular growth morphologies of close space sublimated CdTe films. Nanoscale Advances. 2020;**2**:4757-4769

[121] Flores M, Puente C, Galvan G, Guillen A. CdTe thin films grown by pulsed laser deposition using powder as target: Effect of substrate temperature. Journal of Crystal Growth. 2014;**386**:27-31

[122] Topaz Solar Farm, From Wikipedia. Available from: https://en.wikipedia.org/wiki/Topaz_Solar_Farm [Accessed: December 15, 2021]

[123] Gloeckler M, Zhao Z, Saknkin I. CdTe Solar Cells at the Threshold to 20% Efficiency. IEEE Journal of Photovoltaics. 2013;**3**:1389-1393

[124] Desert Sunlight Solar Farm, from Wikipedia. Available from: https://en.wikipedia.org/wiki/Desert_Sunlight_Solar_Farm [Accessed: December 15, 2021]

[125] Kamala K, Ebin B, Indra S, Bista S, Rijal S, Manoj K. Semi-transparent p-type barium copper sulfide as a back contact interface layer for cadmium telluride solar cells. Solar Energy Materials and Solar Cells. 2020;**218**. DOI: 10.1016/j.solmat.2020.110764

[126] Waldpolenz Solar Park. From Wikipedia. Available from: https://en.wikipedia.org/wiki/Waldpolenz_Solar_Park [Accessed: December 15, 2021]

[127] Devendra K, Deb K, Yang B, Kim Y, Pant B, Park M. Numerical investigation of graphene as a back surface field layer on the performance of cadmium telluride solar cell. Molecules. 2021;**26**. DOI: 10.3390/molecules26113275

[128] Templin Solar Photovoltaic Power plant, Brandemburg. Available from: https://www.renewable-technology.com/projects/templin-solar-photovoltaic-power-plant-brandenburg/ [accessed: December 14, 2021]

[129] Deb K, Amer M, Akhtar M, Devendra K. Impact of different antireflection layers on cadmium telluride (CdTe) solar cells: a PC1D simulation study. Journal of Electronic Materials. 2021;**50**:2199-2205

[130] Xixing W, Lu T, Morris W, Wang G, Bhat I, Jian S. Epitaxial CdTe thin films on mica by vapor transport deposition for flexible solar cells. ACS Applied Energy Materials. 2020;**3**:4589-4599

[131] Alamri N, Alsadi M. Growth of $Cu(In,Ga)Se_2$ thin films by a novel single flash thermal evaporation source. Journal of Taibah University for Science. 2020;**14**:38-43

[132] Kotbi A, Fadili S, Ridah A, Thevenin P, Hartiti B. Synthesis and characterization of sprayed CIGS thin films for photovoltaic application. Materials Today Proceedings. 2020;**24**:66-70

[133] Ying L, Kong D, Li J, Zhao C, Chen C. Preparation of $Cu(In,Ga)Se_2$ thin film by solvothermal and

spin-coating process. Energy Procedia. 2012;**16**:217-222

[134] Negami T, Takuya S, Yasuhiro H, Hayashi S. Large-area CIGS absorbers prepared by physical vapor deposition. Solar Energy Materials and Solar Cells. 2001;**67**:1-9

[135] Lee C. CIGS Thin Film Solar Cells by Electrodeposition. Journal of the Korean Electrochemical Society. 2011;**14**:61-70

[136] Barman B, Kalita P. Influence of back surface field layer on enhancing the efficiency of CIGS solar cell. Solar Energy. 2021;**216**:329-337

[137] Sobayel M, Hossain T, Rashid M, Techato K, Islam S. Efficiency enhancement of CIGS solar cell by cubic silicon carbide as prospective buffer layer. Solar Energy. 2021;**224**:271-278

[138] Sobayel K, Sopian K, Hasan M, Amin N, Karim R, Dar M, et al. Efficiency enhancement of CIGS solar cell by WS2 as window layer through numerical modelling tool. Solar Energy. 2020;**207**:479-485

[139] Boubakeur M, Aissat A, Vilcot P, Arbia B, Maaref H. Enhancement of the efficiency of ultra-thin CIGS/Si structure for solar cell applications. Superlattices and Microstructures. 2020;**138**. DOI: 10.1016/j.spmi.2019.106377

[140] Nour E, Patane S. Single junction-based thin-film CIGS solar cells optimization with efficiencies approaching 24.5%. Optik. 2020;**218**. DOI: 10.1016/j.ijleo.2020.165240

[141] Bhattacharya N, Batchelor W. Thin-film CuIn1−xGa*x*Se2 photovoltaic cells from solution-based precursor layers. Applied Physics Letters. 1999;**75**. DOI: 10.1063/1.124716

[142] Sale of CIGS solar cell panels expected to reach S1 billion by 2013. Available from: https://www.solar-facts-and-advice.com/CIGS-solar-cell.html [Accessed: December 12, 2021]

[143] Thin-film solar overview: cost, types, application, efficiency Available from: https://www.solarfeeds.com/mag/wiki/thin-film-solar/#Copper_Indium_Diselenide_CIS [Accessed: December 12, 2021]

[144] Copper indium gallium selenide solar cells. From Wikipedia. Available from: https://en.wikipedia.org/wiki/Copper_indium_gallium_selenide_solar_cells#Production [Accessed: December 12, 2021]

[145] Scheer R, Walter T, Lewerenz H, Fearheiley M. $CuInS_2$ based thin film solar cell with 10.2% efficiency. Applied Physics Letters. 1993;**63**. DOI: 10.1063/1.110786

[146] Ho SM. Studies of power conversion efficiency and optical properties of Ni3Pb2S2 thin films. Makara Journal of Science. 2017;**21**:119-124

[147] Rafael J, Jeremy P, Yang C, Vera S, Tonio B, Rupak C. Making record-efficiency SnS solar cells by thermal evaporation and atomic layer deposition. Journal of Visualized Experiments. 2015;**99**. DOI: 10.3791/52705

[148] Vera S, Jeremy R, Lee Y, Sun L, Helen P, Roy G. 3.88% efficient tin sulfide solar cells using congruent thermal evaporation. Advanced Materials. 2014;**26**:7488-7492

[149] Teena M, Ramesh K, Naresh N, Venkatesh R. Architecture of monophase InSe thin film structures for solar cell applications. Solar Energy Materials & Solar Cells. 2017;**166**:190-196

[150] Shinde K, Dubal P, Fulari V, Ghodake G. Morphological modulation of Mn:CdSe thin film and its enhanced electrochemical properties. Journal of Electro Analytical Chemistry. 2014;**727**:179-183

[151] Ju L, Kim Y, Mahesh P, Uma V, Dong L. Fabrication of Cu2SnS3 thin film solar cells using Cu/Sn layered metallic precursors prepared by a sputtering process. Solar Energy. 2017;**145**:27-32

[152] Ayaka K, Araki H, Akiko T, Katagiri H. Annealing temperature dependence of photovoltaic properties of solar cells containing Cu_2SnS_3 thin films produced by co-evaporation. Physica Status Solidi (b). 2015;**252**:1239-1243

[153] Mingrui H, Kim J, Lokhande A. Fabrication of sputtered deposited Cu2SnS3 (CTS) thin film solar cell with power conversion efficiency of 2.39. Journal of Alloys and Compounds. 2017;**701**:901-908

[154] Chen Q, Dou X, Li Z, Chen J, Zhou F, Zhuang S. Study on the photovoltaic property of Cu_4SnS_4 synthesized by mechanochemical process. Optik. 2014;**125**:3217-3220

[155] Kazuo J, Ryoichi K, Tsuyoshi K, Satoru Y, Win SM, Hideaki A, et al. Cu_2ZnSnS_4 type thin film solar cells using abundant materials. Thin Solid Films. 2007;**515**:5997-5999

[156] Hironori K, Kotoe S, Tsukasa W, Hiroyuki S, Tomomi K, Shinsuke M. Development of thin film solar cell based on Cu_2ZnSnS_4 thin films. Solar energy Materials and Solar Cells. 2001;**65**:141-148

[157] Wang K, Gunawan O, Todorov T, Shin B, Chey SJ, Bojarczuk NA, et al. Thermally evaporated Cu_2ZnSnS_4 solar cells. Applied Physics Letters. 2010;**97**. DOI: 10.1063/1.3499284

[158] Schubert B, Marsen B, Cinque S, Unold T, Klenk R, Schorr S, et al. Cu_2ZnSnS_4 thin film solar cells by fast co-evaporation. Progress in Photovoltaics. 2011;**19**:93-96

[159] Chet S, Matthew GP, Vahid A, Brian G, Bonil K, Brian AK. Synthesis of Cu_2ZnSnS_4 nanocrystals for use in low cost photovoltaics. Journal of the American Chemical Society. 2009, 2009;**131**:12554-12555

[160] Jonathan JS, Dominik MB, Philip JD. A 3.2% efficient kesterite device from electrodeposited stacked elemental layers. Journal of Electroanalytical Chemistry. 2010;**646**:52-59

[161] Ennaoui A, Steiner ML, Weber A, Abou-Ras D, Kotschau I, Schock HW, et al. Cu_2ZnSnS_4 thin film solar cells from electroplated precursors: Novel low cost perspective. Thin Solid Films. 2009;**517**:2511-2514

[162] Sawanta SM, Pravin SS, Chirayath AB, Popatrao NB, Young WO, Pramod SP. Synthesis and characterization of Cu_2ZnSnS_4 thin films by SILAR method. Journal of Physics and Chemistry of Solids. 2012;**73**:735-740

[163] Tsukasa W, Tomokazu S, Shin T, Tatsuo F, Tomoyoshi M, Kazuo J, et al. 6% efficiency Cu_2ZnSnS_4 based thin film solar cells using oxide precursors by open atmosphere type CVD. Journal of Materials Chemistry. 2012;**22**:4021-4024

[164] Shinde NM, Dubal DP, Dhawale DS, Lokhande CD, Kim JH, Moon JH. Room temperature novel chemical synthesis of Cu_2ZnSnS_4 (CZTS) absorbing layer

for photovoltaic application. Materials Research Bulletin. 2012;**47**:302-307

[165] Donghyeok S, Kim Y, Hwang D, Kim D, Son C, Park J. Effect of RF power on the properties of sputtered-CuS thin films for photovoltaic applications. Energies. 2020;**13**. DOI: 10.3390/en13030688

[166] Karl W. Cadmium sulfide enhances solar cell efficiency. Energy Conversion and Management. 2011;**52**:426-430

[167] Qiu K, Cai L, Qiu D, Wu W, Liang Z. Preparation of ZnS thin films and ZnS/p-Si heterojunction solar cells. Materials Letters. 2017;**198**:23-26

[168] Omer S, Arzu E, Sabit H. Synthesis of PbS:Mo(3%) thin film and investigation of its properties. Journal of Materials Science: Materials in Electronics. 2019;**30**:7600-7605

[169] Wen X, Chen C, Wang C, Tang J, Niu G, Jun Z. Vapor transport deposition of antimony selenide thin film solar cells with 7.6% efficiency. Nature Communications. 2018;**9**. DOI: 10.1038/s41467-018-04634-6

[170] Liang W, Li D, Chao C, Niu G, Tang J, Fan J, et al. Stable 6% efficient Sb_2Se_3 solar cells with a ZnO buffer layer. Nature Energy. 2017;**2**. DOI: 10.1038/nenergy.2017.46

[171] Chao C, Li W, Ying Z, Chen C, Liu X, Bo Y, et al. Optical properties of amorphous and polycrystalline Sb_2Se_3 thin films prepared by thermal evaporation. Applied Physics Letters. 2015;**107**. DOI: 10.1063/1.4927741

[172] Manorama L, Mahapatra K, Nandu B. Development of $CuInTe_2$ thin film solar cells by electrochemical route with low temperature (80 °C) heat treatment procedure. Materials Science and Engineering: B. 2016;**204**:20-26

[173] Lakhe M, Nandu B. Characterization of electrochemically deposited $CuInTe_2$ thin films for solar cell applications. Solar Energy Materials and Solar Cells. 2014;**123**:122-129

[174] Jia G, Liu J, Zhang W, Li R, Kun W, Yang P, et al. $CuInTe_2$ nanocrystals: Shape and size control, formation mechanism and application, and use as photovoltaics. Nanomaterials. 2019;**9**. DOI: 10.3390/nano9030409

[175] Hyun Y, Ji H, Joshi B, Ra M, Sam S, Yoon K, et al. $CuInSe_2$ (CIS) thin film solar cells by electrostatic spray deposition. Journal of The Electrochemical Society. 2012;**159**. DOI: 10.1149/2.jes113086

[176] Se J, Kim C, Yun H, Gwak J, Jeong S, Ryu B, et al. $CuInSe_2$ (CIS) thin film solar cells by direct coating and selenization of solution precursors. The Journal of Physical Chemistry C. 2010;**114**:8108-8113

[177] Prabukanthan P, Lakshmi R, Tetiana T. Photovoltaic device performance of pure, manganese (Mn^{2+}) doped and irradiated $CuInSe_2$ thin films. New Journal of Chemistry. 2018;**42**:11642-11652

Chapter 2

Contribution to the Calculation of Physical Properties of BeSe Semiconductor

Mohamed Amine Ghebouli and Brahim Ghebouli

Abstract

We expose various physical parameters of binary compound BeSe in the stable zinc blend and NiAs structures using the functional HSE hybrid, GGA-PBE, and LDA. We deduce elastic constants, mechanical parameters, and wave velocities according to different orientations. BeSe semiconductor has Γ-X (2.852 eV) and Γ-K (0.536 eV) bandgap in zinc blend and NiAs structures. Electrons transit from Se-p site to the Be-s state and show covalent bonding. Optical absorption peaks result from electronic transitions under ultraviolet light irradiation.

Keywords: CASTEP, zinc blend, NiAs, band structure, bandgap, absorption, BeSe

1. Introduction

Researchers have focused on II-VI semiconductors, which are intended for technological applications. BeSe crystallizes in zinc blend structure at room temperature and its properties depend on structure, pressure, temperature, and calculation method. The HSE hybrid underestimates the lattice parameters compared to GGA-PBE and LDA but gives a bandgap in accordance with the experimental value. The phonon dispersion curve of BeSe shows two optical and acoustic couplets and confirms its stability. The maximum of reflectivity (63% and 56%) is observed in the ultraviolet light domain. We expose the detailed calculation concerning the lattice parameters, elastic constants, bandgap, phonons frequencies, and optical parameters. For works carried out by other researchers, we note the transition from zinc blend phase to NiAs structure at 55 GPa [1]. The study of elastic constants and bandgap of BeSe in zinc blend phase [2, 3], the experimental bandgap ranging from 2 eV to 4.5 eV [4, 5], a small ionic radius ratio, a high degree of covalent bonding, and high hardness [6]. Appropriate bandgap and optical absorption parameters make BeSe as an important material in the area of catalysis and luminescent devices [7–9].

2. Computational method

We perform calculations on BeSe by using the CASTEP code [10]. The energy E_{cut} of 660 eV and k-points of $8 \times 8 \times 8$ using Monkhorst-Pack scheme [11] ensure well convergence. We use 20 × 20 × 20 k-points in the computation of optical parameters. The GGA, LDA, and HSE hybrid functional [12, 13] treat the exchange-correlation potential. Broyden-Fletcher-Goldfarb-Shanno (BFGS) minimization technique [14] determines the structural parameters. The tolerance of geometry optimization was a total energy 5×10^{-6} eV/atom, maximum ionic Hellmann-Feynman force 0.01 eV/Å, and maximum stress 0.02 eV/Å^3.

3. Results and discussions

3.1 Ground state

The position of Be atom is the corner (0, 0, 0), while the Se atom takes place at (1/4, 1/4, 1/4) and (1/3, 2/3, 1/4) for zinc blend and NiAs phases. **Figure 1(a)** gives

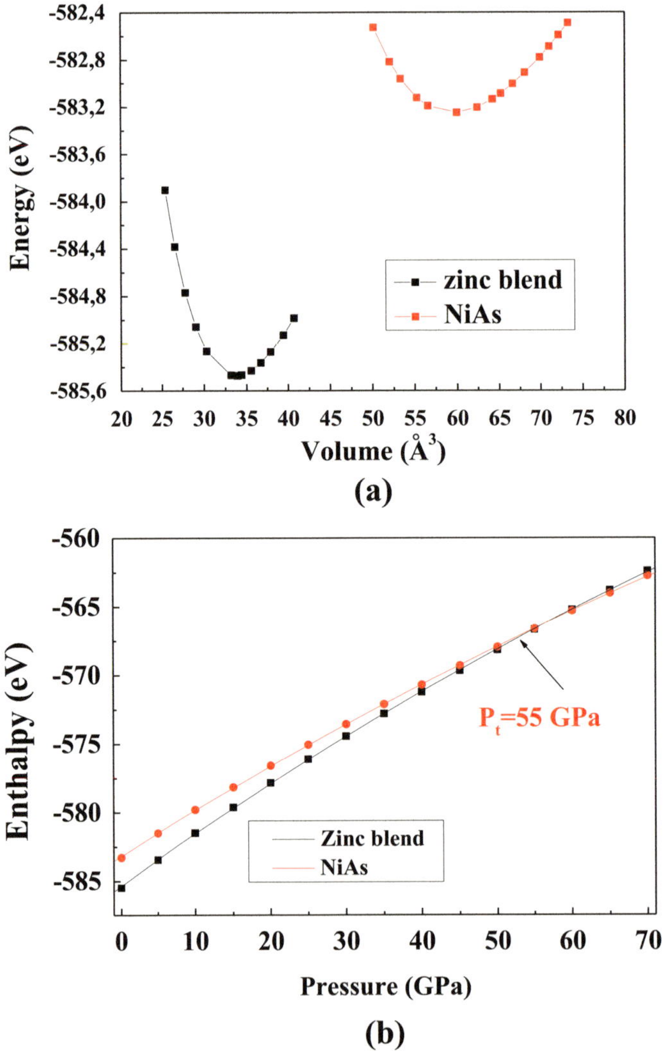

Figure 1.
The plots of total energy versus volume (a) and enthalpy versus pressure (b) for BeSe in zinc blend and NiAs phases.

	Zinc blend					NiAs				
	GGA	LDA	HSE hybrid	Exp	Other	GGA	LDA	HSE hybrid	Exp	Other
a (Å)	5.129	5.083	4.976	5.139 [15]	5.137 [16] 5.037 [17] 5.087 [18]	3.513	3.456	3.415		3.524 [16] 3.421 [17] 3.502 [19]
c (Å)						5.459	5.461	5.347		5.462 [16] 5.398 [17] 5.529 [19]
B_0 (GPa)	74.50	87.35	99.484	92.20 [15]	74.97 [18] 80 [4] 96.5 [20]	89.20	100.30	112.76		94.7 [17] 86 [21] 109.6 [22]
B'	3.82	3.65	3.565		4.02 [23] 3.58 [24] 3.85 [19]	3.94	3.826	3.749		3.852 [17] 3.70 [25]

Table 1.
Lattice parameters, bulk modulus, and its pressure derivative for BeSe in zinc blend and NiAs structures within GGA-PBE, LDA, and HSE hybrid functional.

the plots of total energy versus unit cell volume, where the zinc blend phase is more stable. **Figure 1(b)** visualizes the effect of pressure on formation enthalpy. The two enthalpy-pressure diagrams show a transition region around 55 GPa [1]. We report in **Table 1** that the lattice parameters, bulk modulus, and its pressure derivative are calculated with GGA, LDA, and HSE hybrid. Our results agree well with experimental measurements [15] and theoretical data [4, 16–25]. The HSE hybrid underestimates the lattice parameters.

3.2 Elastic constants and mechanical parameters

Table 2 includes elastic moduli of BeSe in both phases within GGA and LDA. Our elastic constants of zinc blend structure agree with theoretical values [18, 22, 26, 27]. The following relationships [28, 29] ensure the elastic stability of BeSe.

Zinc blend

$$0\langle C_{11} + 2C_{12}, 0\langle C_{44}, 0\langle C_{11} - C_{12}, C_{12}\langle B\langle C_{11} \quad (1)$$

NiAs

$$0\langle C_{11} - |C_{12}|, 0\langle C_{33}(C_{11} + C_{12}) - 2C_{13}^2, 0\langle C_{44} \quad (2)$$

Figure 2 shows the dependence of elastic moduli on pressure for beryllium selenide in zinc blend and NiAs phases. We list in **Table 2** the bulk modulus, shear modulus, Young's modulus, Poisson's ratio, anisotropy factor, and B_H/G_H ratio. The Poisson's ratio and anisotropy factor indicate the covalent bonding and strong anisotropy in BeSe. The reported B_H/G_H value translates the brittle zinc blend phase and ductile NiAs structure. **Figure 3** visualizes the effect of orientation on Poisson's ratio (a) and shear modulus (b) in xy, xz, and yz planes. It can be seen that the two parameters are isotropic only in the B8 phase and in the xy plane. **Table 3** predicts the maximum and minimum values of Young's modulus, linear compressibility, shear

		Zinc blend			NiAs	
		GGA	LDA	Other	GGA	LDA
C_{11} (GPa)		120.06	137.22	117 [18] 145 [24]	140.27	167.04
C_{12} (GPa)		30.73	55.94	45.5 [26] 51 [22]	71.51	80.20
C_{44} (GPa)		66.2	74.35	73.39 [26] 61 [22]	8.48	14.57
C_{13} (GPa)					36.45	46.51
C_{33} (GPa)					178.40	202.79
Θ_D (K)		460.91	465.77	407 [26]	280.75	332.74
B(GPa)	B_V	60.51	83.08		83.08	98.15
	B_R	60.51	83.08		83.08	98.15
	B_H	60.51	83.08		83.08	98.14
G (GPa)	G_V	131.16	60.86	66.24 [27]	31.23	60.86
	G_R	127.51	55.83	61.14 [27]	16.22	55.83
	G_H	129.34	58.53	63.78 [27]	23.73	58.53
E_H (GPa)		129.34	141.82	153.42 [27]	65.00	86.89
σ_H		0.138	0.21	0.203 [27]	0.36	0.35
A^U		0.188	0.45		4.62	2.60
B_H/G_H		0.46	1.41	1.349 [27]	3.5	3.05

Table 2.
Elastic constants C_{ij}, Debye temperature Θ_D, bulk modulus B, shear modulus G, Young's modulus E_H, Poisson's ratio σ_H, anisotropy factor A^U and ratio B_H/G_H for BeSe in zinc blend and NiAs structures.

modulus, and Poisson's ratio. **Table 2** reports the Debye temperature of BeSe, which agrees with the theoretical value [26]. **Table 4** predicts elastic wave velocities along [100], [110], [111], and [001] directions.

3.3 Electronic parameters

Figure 4 shows the band structure of BeSe using GGA-PBE and HSE hybrid functional. We observe the indirect bandgap Γ-X (Γ-K) in the zinc blend (NiAs) structure. **Table 5** reports the bandgap between various symmetry points. The bandgap of BeSe in the zinc blend structure agrees with other calculations [3, 16, 23, 31]. **Figure 5** displays the dependence on the pressure of the direct and indirect bandgap. The bandgap calculated using the HSE hybrid agrees with the experimental one [3, 32, 33]. **Figure 6** shows the plots of total and partial densities of states. Se-p and Be-s orbitals are responses for the upper valence band. The first conduction band is wider for both structures and consists of Be-s and Se-p orbitals, with a small contribution of the Se-s site. The optical transition occurs from Se-p state to the Be-s site. There is a hybridization between Se-p and Be-s in the upper valence band, which translates their covalent bonding.

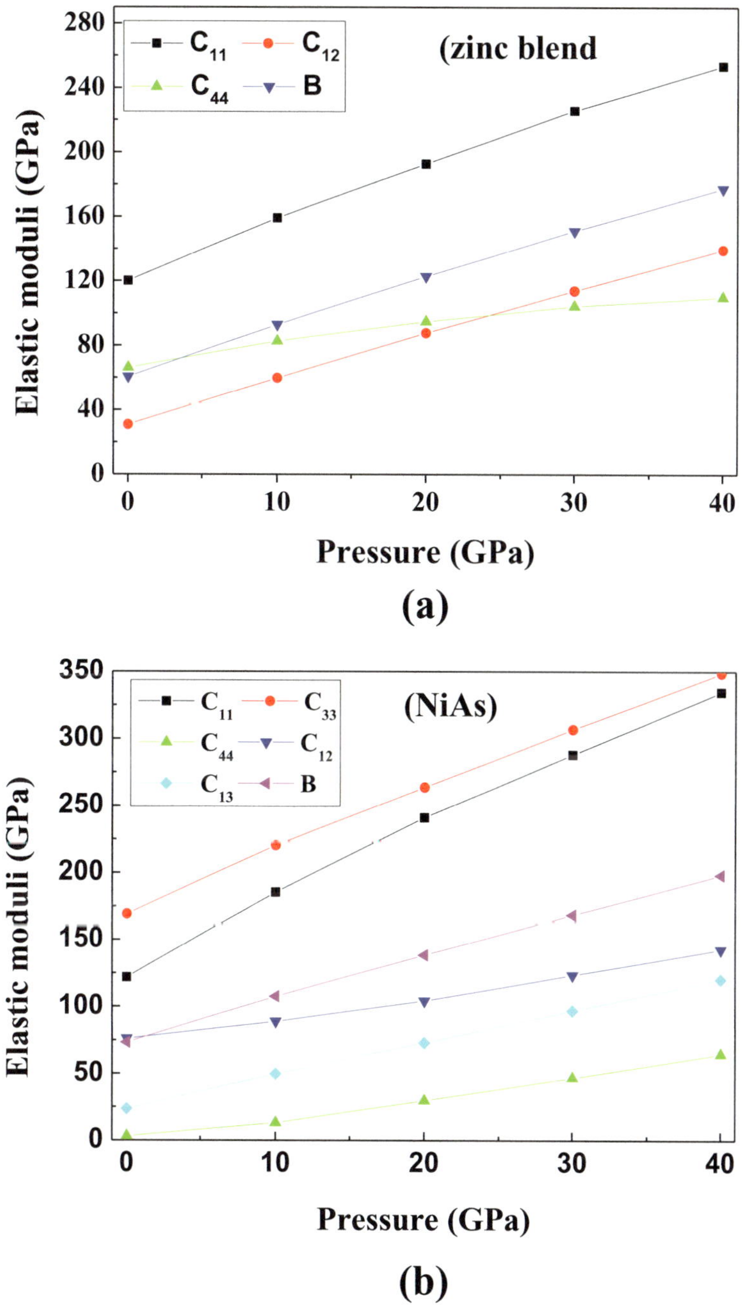

Figure 2.
Elastic moduli versus pressure in zinc blend (a) and NiAs (b) phases.

3.4 Phonon frequencies

Figure 7 shows the phonon dispersion curve of BeSe in zinc blend structure. Optical (acoustic) phonons exhibit two couplets of longitudinal optical (LO)

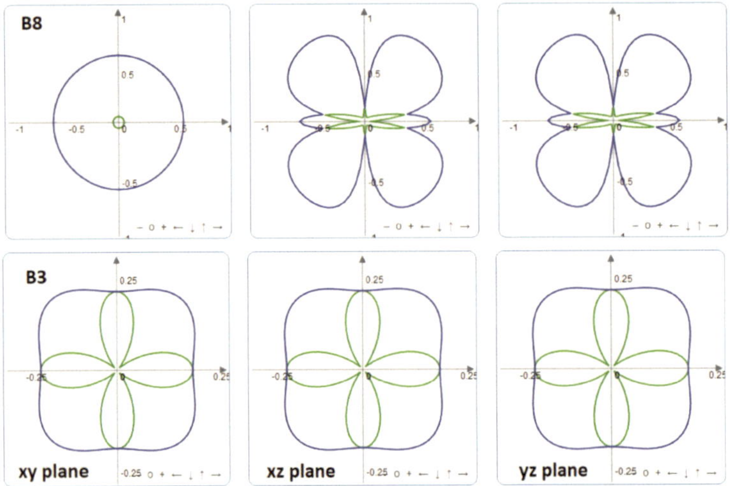

(a)

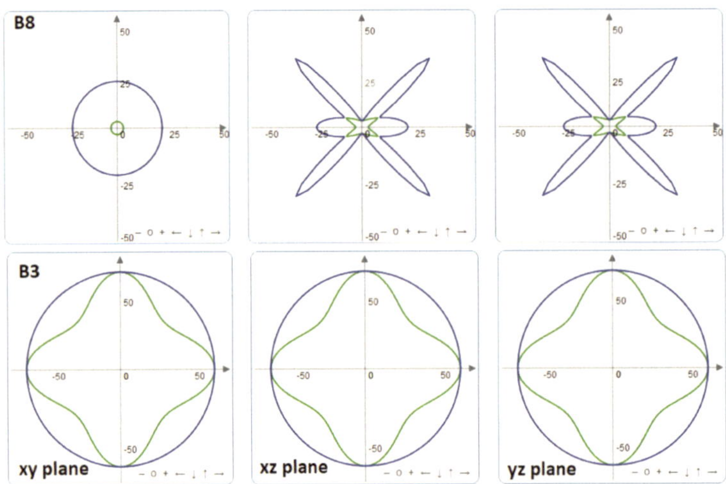

(b)

Figure 3.
The orientation effect on Poisson's ratio (a) and shear modulus (b) in zinc blend and NiAs phases.

Parameters		Young's modulus		Linear compressibility		Shear modulus		Poisson's ratio	
		E_{min}	E_{max}	β_{min}	β_{max}	G_{min}	G_{max}	σ_{min}	σ_{max}
Zinc blend	GGA	107.54	145.55	5.5083	5.5083	44.665	44.665	0.009	0.253
	LDA	104.83	171.79	4.0142	4.0142	40.644	74.354	0.0039	0.409
NiAs	GGA	42.108	163.56	4.5012	4.5012	13.201	47.33	0.118	0.623
	LDA	49.954	185.29	3.3662	3.3662	14.575	64.941	0.102	0.719

Table 3.
The maximum and minimum values of Young's modulus E, linear compressibility β, shear modulus G, and Poisson's ratio σ for BeSe in zinc blend and NiAs structures.

Material	Directions	[100]	[110]	[111]	[001]
Zinc blend	v_l (ms^{-1})	5439	5898	6043	—
	v_{t1} (ms^{-1})	3803	4305	3316	—
	v_{t2} (ms^{-1})	3803	3803	3316	—
NiAs	v_l (ms^{-1})	5192	—		6067
	v_{t1} (ms^{-1})	2187	—		1243
	v_{t2} (ms^{-1})	1243	—		1243

Table 4.
Sound velocities for BeSe along with main directions in zinc blend and NiAs structures.

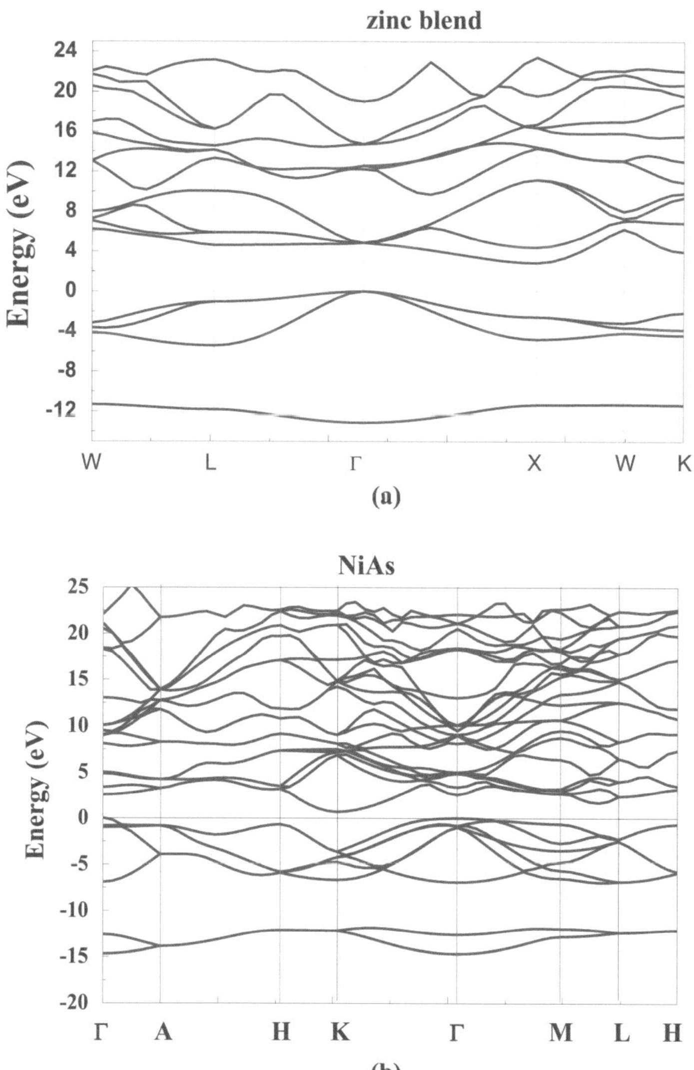

Figure 4.
The band structure of BeSe in zinc blend (a) and NiAs.(b) phases at various points in the Brillouin zone.

	Zinc blend					NiAs		
	GGA	**LDA**	**HSE hybrid**	**Exp**	**Other**	**GGA**	**LDA**	**HSE hybrid**
$E_{\Gamma\text{-}\Gamma}$ (eV)	4.84	4.42	5.81	4.72 [3] 5.5 [30]		2.55	2.43	3.45
$E_{L\text{-}L}$ (eV)	5.65	5.52	7.40			4.64	4.88	6.02
$E_{X\text{-}X}$ (eV)	5.41	5.20	6.71					
$E_{\Gamma\text{-}X}$ (eV)	2.85	2.408	3.45	4–4.5 [3]	2.41 [31] 3.12 [16]			
$E_{\Gamma\text{-}L}$ (eV)	4.60	4.39	6.05		4.33 [23]	2.38	2.27	6.23
$E_{\Gamma\text{-}K}$ (eV)	3.94	3.62	4.71	4.73 [32]		0.68	0.51	1.16
$E_{M\text{-}M}$ (eV)						3.28	2.98	3.94

Table 5.
Band gaps for BeSe at equilibrium between various symmetry points in zinc blend and NiAs structures.

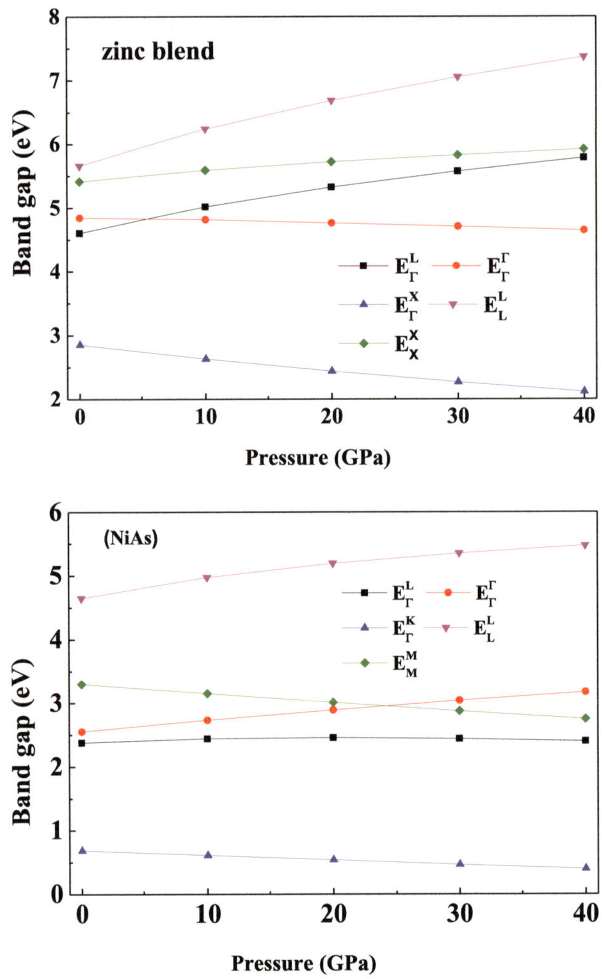

Figure 5.
The band gaps versus pressure in zinc blend and NiAs phases.

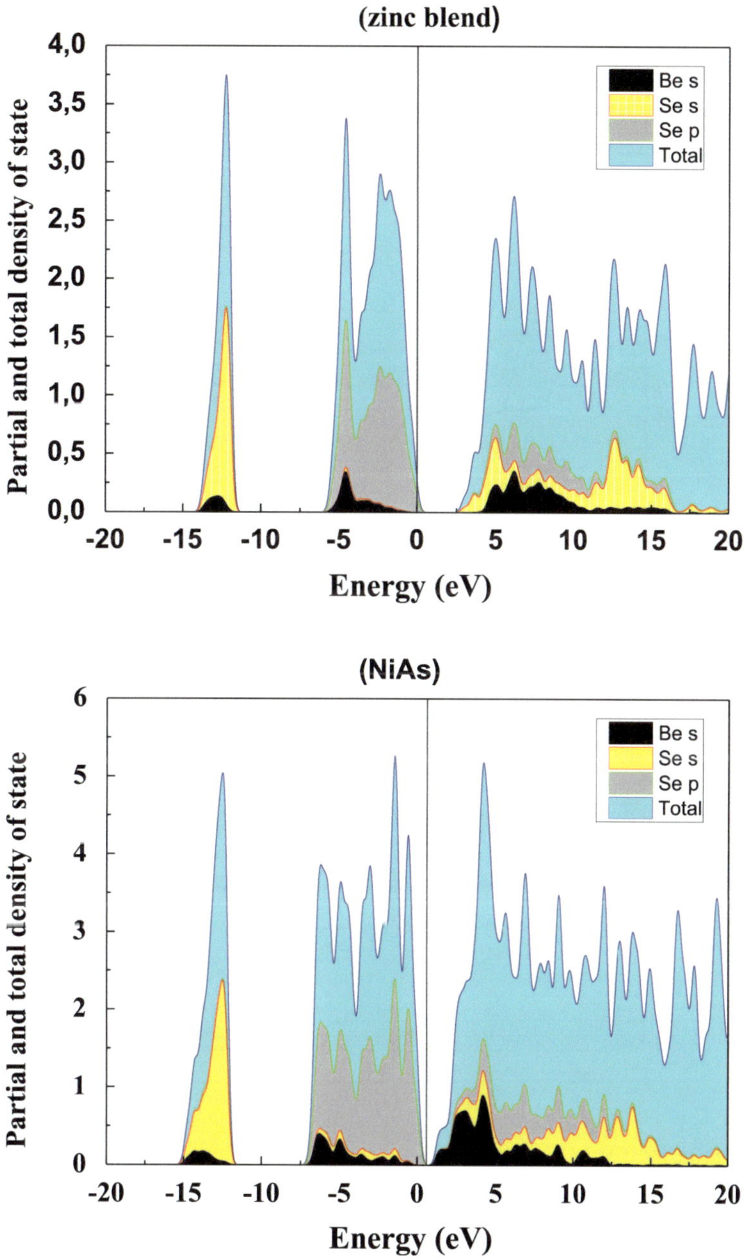

Figure 6.
Total and partial densities of states.

{longitudinal acoustic (LA)} and transverse optical (TO) {transverse acoustic (TA)} modes and confirm the dynamical stability of BeSe. The frequency 16.06 cm^{-1} (198 cm^{-1}) separates optical (acoustic) longitudinal and transversal branches. The maximum of longitudinal optical phonons is at X point. **Figure 8** shows the effect of pressure on phonon frequencies ω_{LO} and ω_{TO} of BeSe. The equilibrium values of ω_{LO} and ω_{TO} at Γ point are 537 cm^{-1} and 471 cm^{-1}.

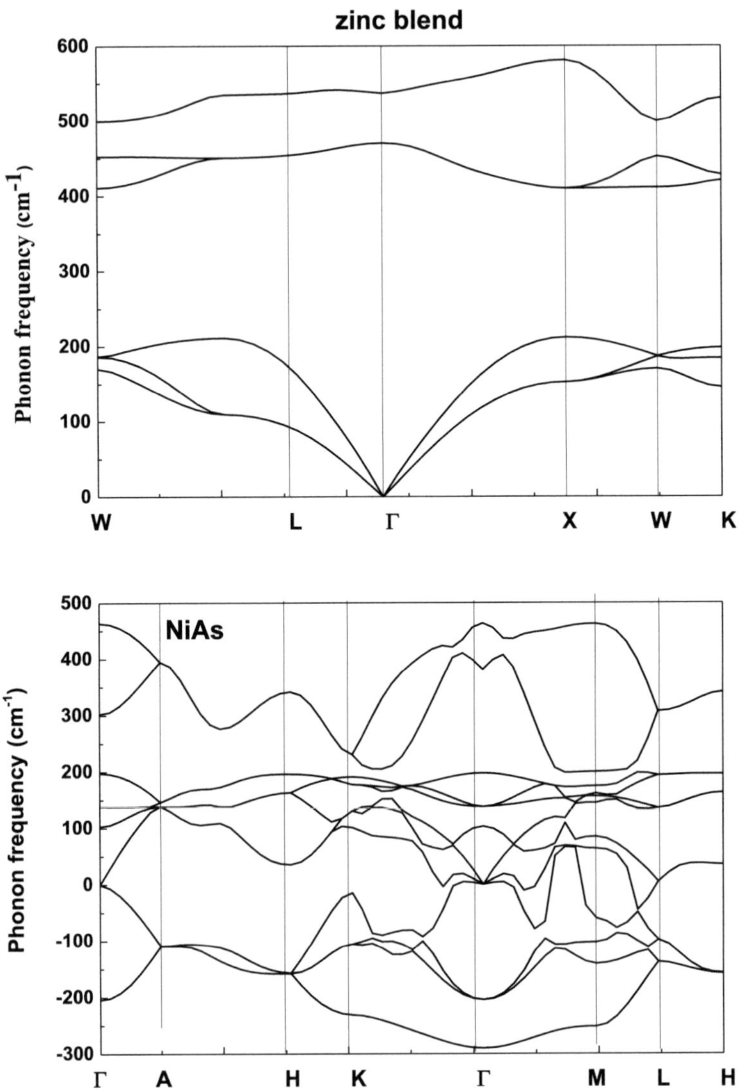

Figure 7.
The phonon dispersion curves.

3.5 Optical parameters

3.5.1 Loss function

We plot in **Figure 9** the loss function spectra versus wavelength for BeSe in zinc blend and NiAs structures. The high-loss region has a wavelength ranging from 42 nm to 90 nm. The low-loss function has a wavelength range less than 40 nm and greater than 90 nm. The maximum loss function reaches the value of 4.84% (4.49%) at 60.32 nm (57.25 nm) in the zinc blend (NiAs) structure. There is no loss in the ultraviolet and visible light domain.

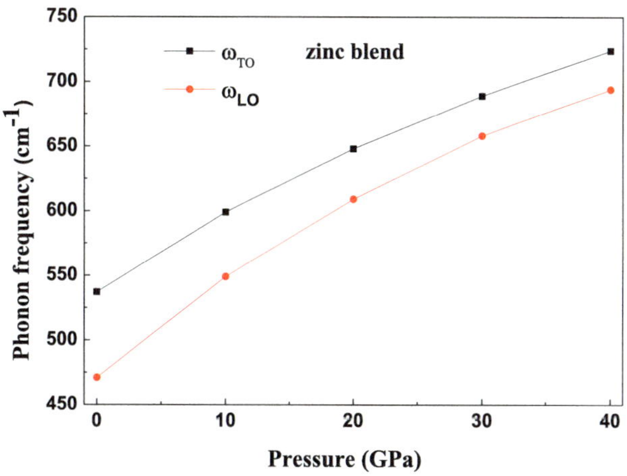

Figure 8.
The phonon frequency versus pressure at Γ point.

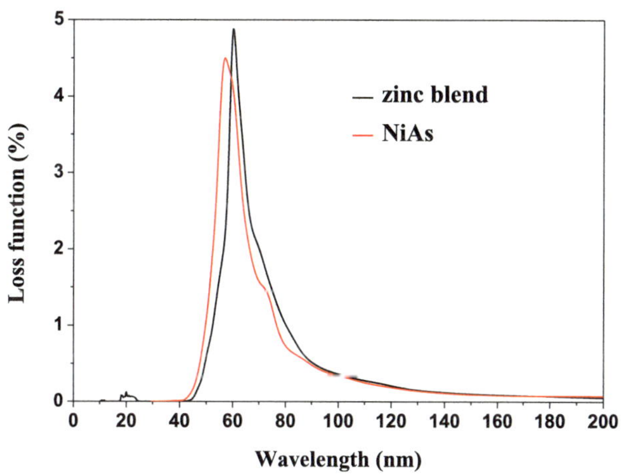

Figure 9.
The loss function versus wavelength.

3.5.2 Optical absorption

We applied a smearing value of 0.1 to obtain more distinguishable absorbance peaks. We attribute the absorption peaks as depicted in **Figure 10** to the photo transition energies from the maximum valence band to the minimum conduction band under ultraviolet light irradiation. BeSe absorbs a maximum of ultraviolet light 331,971 cm^{-1} (335,643 cm^{-1}) at wavelength 163.6 nm (163 nm) in zinc blend (NiAs) structure, shows narrow indirect bandgap, then it is a good candidate for photocatalysis in the ultraviolet light domain (120–400 nm).

3.5.3 Optical reflectivity

The reflectivity estimates the amount of incident light on the surface of photocatalytic material. The reflectivity of BeSe, as shown in **Figure 11**, starts at 0.007

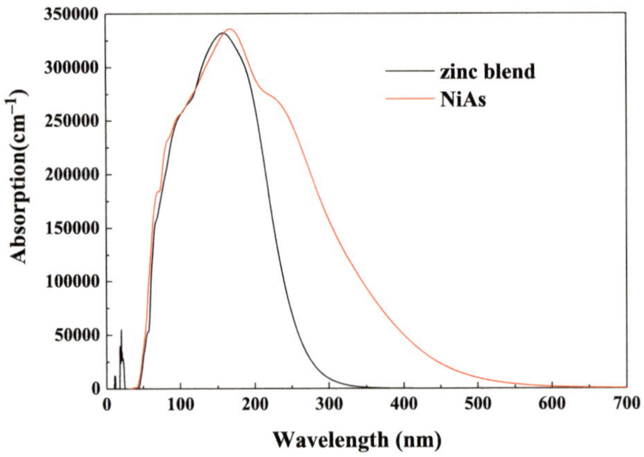

Figure 10.
The absorption versus wavelength.

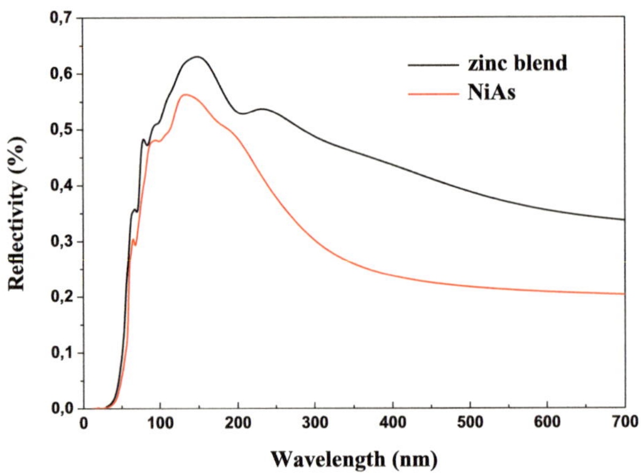

Figure 11.
The reflectivity versus wavelength.

and corresponds to a wavelength around 33 nm. It reaches a maximum value 0.63% (0.56%) at 149 nm (136.5 nm) in zinc blend (NiAs) structure. The reflectivity in the ultraviolet light domain (120 nm–400 nm) is 0.63% (0.56%) in zinc blend (NiAs) structure. While in the visible light (400 nm–800 nm), it is 0.43% (0.23%) for zinc blend (NiAs) phase.

4. Conclusion

We study BeSe in zinc blend and NiAs phases using GGA-PBE, LDA, and HSE hybrid. BeSe is a good candidate for photocatalysis material. We estimate the lattice parameters, shear modulus, Young's modulus, Poisson's ratio, average sound velocities, and Debye temperature. The maximum absorption at ultraviolet light is 331,971 cm^{-1} (335,643 cm^{-1}) in zinc blend (NiAs) structure. We perform transverse and longitudinal optical phonon frequencies ω_{TO} and ω_{LO} at Γ point. A frequency gap of 16.06 cm^{-1} (198 cm^{-1}) in zinc blend structure separates longitudinal and transversal optical (acoustic) branches. The reflectivity in the ultraviolet light domain is 0.63% (0.56%) in zinc blend (NiAs) structure. While in the visible light, the reflectivity is 0.43% (0.23%) for the zinc blend (NiAs) phase.

Author details

Mohamed Amine Ghebouli[1]* and Brahim Ghebouli[2]

1 Faculty of Technology, Department of Chemistry, University of Mohamed Boudiaf, M'sila, Algeria

2 Laboratory of Studies Surfaces and Interfaces of Solids Materials, Faculty of Science, Department of Physics, University Ferhat Abbas of Setif 1, Algeria

*Address all correspondence to: mohamedamine.ghebouli@univ-msila.dz

References

[1] Ghebouli MA, Ghebouli B, Chihi T, Fatmi M. Study of structural, elastic, electronic, dynamical and optical properties of beryllium selenide (BeSe) semiconductor in zinc blend and NiAs phases. Physica B Condensed Matter. 2021;**610**(1):412858. DOI: 10.1016/j.physb.2021.412858

[2] González-Díaz M, Rodríguez-Hernández P, Munõz A. Elastic constants and electronic structure of beryllium chalcogenides BeS, BeSe, and BeTefrom first-principles calculations. Physical Review B. 1997;**55**(21):14043

[3] Srivastava GP, Tütüncü HM, Günhan N. First-principles studies of structural, electronic, and dynamical properties of Be chalcogenides. Physical Review B. 2004;**70**:085206

[4] Yim W, Dismukes J, Stofko E, Paff R. Synthesis and some properties of BeTe, BeSe and BeS. Journal of Physics and Chemistry of Solids. 1972;**33**:501-505

[5] Bouhafs B, Aourag H, Ferhat M, Certier M. Competition between the ionic and covalent character in the series of boron compounds BP, BAs, and BSb. Phys J. Condensed Matter. 1999;**11**(30):5781

[6] Khenata R, Bouhemadou A, Hichour M, Baltache H, Rached D, Rérat M. Elastic and optical properties of BeS, BeSe and BeTe under pressure. Solid State Electronics. 2006;**50**(7–8):1382-1388

[7] Pandey R, Sivaraman S. Spectroscopic properties of defects in alkaline-earth sulfides. Journal of Physics and Chemistry of Solids. 1991;**52**(1):211-225

[8] Asano S, Yamashita N, Nakao Y. Luminescence of the Pb^{2+}-Ion Dimer Center in CaS and CaSe Phosphors. Physica Status Solidi. 1978;**89**(2):663-673

[9] Nakanishi Y, Ito T, Hatanaka Y, Shimaoka G. Preparation and luminescent properties of SrSe: Ce thin films. Applied Surface Science. 1992;**66**:515

[10] Segall MD, Lindan PJD, Probert MJ, Pickard CJ, Hasnip PJ, Clark SJ, et al. First-principles simulation: Ideas, illustrations and the CASTEP code. Journal of Physics: Condensed Matter. 2002;**14**(11):2717-2744

[11] Monkhorst HJ, Pack JD. Special points for Brillouin-zone integrations. Physical Review B. 1976;**13**(12):5188

[12] Perdew JP, Burke K, Ernzerhof M. Generalized gradient approximation made simple. Physical Review Letters. 1996;**77**(18):3865

[13] Goedecker S, Teter M, Hutter J. Separable dual-space Gaussian pseudopotentials. Physical Review B. 1996;**54**(3):1703

[14] Fischer TH, Almlof J. General methods for geometry and wave function optimization. The Journal of Physical Chemistry. 1992;**96**(24):9768-9774

[15] Luo H, Ghandehari K, Greene RG, Ruoff AL. Phase transformation of BeSe and BeTe to the NiAs structure at high pressure. Physical Review B. 1995;**52**(10):7058

[16] Kalpana G, Pari G, Mookerjee A, Bhattacharyya AK. *Ab initio* Electronic Band Structure Calculations for Beryllium Chalcogenides. International Journal of Modern Physics B. 1998;**12**(19):1975

[17] Munõz A, Rodriguez-Hernandez P, Mujica A. Ground-state properties and high-pressure phase of beryllium chalcogenides BeSe, BeTe, and BeS. Physical Review B. 1996;**54**(17):11861

[18] Heciri D, Beldi L, Drablia S, Meradji H, Derradji NE, Belkhir H, et al. First-principles elastic constants and electronic structure of beryllium chalcogenides BeS, BeSe and BeTe. Computational Materials Science. 2007;**38**(4):609-617

[19] Bouamama K, Daoud K, Kassali K. Ab initio calculations in the virtual-crystal approximation of the structural and the elastic properties of BeSxSe1–x alloys under high pressure. Modelling Simul. Materials Science and Engineering. 2005;**13**(7):1153

[20] Dutta R, Alptekin S, Mandal N. Electronic structure, optical properties and the mechanism of the B3–B8 phase transition of BeSe: Insights from hybrid functionals, lattice dynamics and *NPH* molecular dynamics. Journal of Physics: Condensed Matter. 2013;**25**(12):125401

[21] Munjal N, Sharma V, Sharma G, Vyas V, Sharma BK, Lowther JE. Ab-initio study of the electronic and elastic properties of beryllium chalcogenides BeX (X= S, Se and Te). Physica Scripta. 2011;**84**(3):035704

[22] Berghout A, Zaoui A, Hugel J. Fundamental state quantities and high-pressure phase transition in beryllium chalcogenides. Journal of Physics: Condensed Matter. 2006;**18**(46): 10365-10375

[23] Okoye CMI. Structural, electronic, and optical properties of beryllium monochalcogenides. European Physical Journal B. 2004;**39**(1):5

[24] Postnikov AV, Pages O, Tite T, Ajjoun M, Hugel J. Examination of Size-Induced Ferroelectric Phase Transitions in Template Synthesized PbTiO3 Nanotubes and Nanofibers. Phase Transitions. 2005;**78**(9–11):219

[25] Hassan FEH, Akbarzadeh H. Ground state properties and structural phase transition of beryllium chalcogenides. Computational Materials Science. 2006; **35**(4):423-431

[26] Kong F, Jiang G. Phase transition, elastic, thermodynamic properties of zinc-blend BeSe from first-principles. Physica B. 2009;**404**(21):3935-3940

[27] Chattopadhyaya S, Sarkar U, Debnath B, Debbarma M, Ghosh D, Chanda S, et al. Density functional study of structural, elastic, electronic and optical properties of BexCd1–xS, BexCd1–xSe and BexCd1–xTe alloys using FPLAPW approach. Physica B: Condensed Matter. 2019;**563**:1-22. DOI: 10.1016/j.physb.2019.03.025

[28] Cruz WDL, Duaz JA, Mancera L, Takeuchi N, Soto G. Yttrium nitride thin films grown by reactive laser ablation. Journal of Physics and Chemistry of Solids. 2003;**64**:2273

[29] Born M, Hang K. Dynamical Theory and Experiments I. Berlin: Springer Verlag Publishers; 1982

[30] Berning PH, Hass G, Madden RP. Reflectance-increasing coatings for the vacuum ultraviolet and their applications. Journal of the Optical Society of America. 1960;**50**(6):586-597

[31] Rached D, Rabah M, Benkhettou N, Khenata R, Soudini B, Al-Douri Y, et al. First-principle study of structural, electronic and elastic properties of beryllium chalcogenides BeS, BeSe and BeTe. Computational Materials Science. 2006;**37**(3):292-299

[32] Sarkar RL, Chatterjee S. Electronic energy bands of BeS, BeSe and BeTe. Journal of Physics C: Solid State Physics. 1977;**10**(1):57-62. DOI: 10.1088/0022-3719/10/1/011

[33] Wilmers K, Wethkamp T, Essar N, Cobet C, Richter F, Wagner V, et al. Ellipsometric studies of BexZn1–xSe between 3 eV and 25 eV. Physical Review B. 1999;**59**(15):10071-10075

Chapter 3

Thickness Dependent Spectroscopic Studies in 2D $PtSe_2$

Nilanjan Basu, Vishal K. Pathak, Laxman Gilua and Pramoda K. Nayak

Abstract

Transition metal dichalcogenides (TMDCs) are emerging to be an exciting class of 2D materials apart from graphene or hexagonal boron nitride (*h*-BN). They are a class of layered materials that exhibit inspiring properties which are worth exploring, among them $PtSe_2$ is fairly a new addition. Although bulk $PtSe_2$ was first synthesized more than a century ago, the study of its layer-dependent properties is still at a nascent stage. The monolayer of $PtSe_2$ exhibits a band gap between 1.2 and 1.8 eV, the band gap starts to decrease with an increase in the number of layers thus transforming into semimetal type. Among all other 2D materials it shows the highest electron mobility of about 3000 $cm^2\ V^{-1}\ s^{-1}$ and unlike other TMDCs, it is strikingly stable in ambient conditions. Owing to its stability and tunable properties, it has great potential in the fields of optoelectronics, spintronics, sensorics, and many more. In this book chapter, we report the thickness dependent spectroscopic properties of mechanically exfoliated $PtSe_2$. We have explored low temperature Raman spectroscopy as well as polarized Raman spectroscopy to study in detail the vibrational properties of $PtSe_2$. Raman spectroscopy is also employed to determine its thermal conductivity. We hope that this work will provide a fresh overview of $PtSe_2$ from a spectroscopic perspective.

Keywords: transition metal dichalcogenides, $PtSe_2$, raman spectroscopy, electronic band structure, thermal conductivity

1. Introduction

Once a forbidden material, one atom thick material came into existence with the discovery of graphene in 2004 by Novoselov et al. [1]. Unsurprisingly due to the unique properties and momentous potential of 2D materials, the research world jumped into the foray. Interestingly the term "graphene" was coined in 1986 by Boehm and the electronic band structure of single layer graphite is studied since 1942 [2]. The surge of discovery didn't stop with graphene, it gathered pace, and "2D materials" like *h*-BN, transition metal dichalcogenides (TMDCs), and several other layered materials came into existence. This present work deals with one of the less explored TMDCs, i.e. $PtSe_2$. Its unique and exciting properties are worth exploring. It has a wide range of applications like photodetectors, gas sensing, electronics, piezoresistive sensors, and electrocatalysis [3–6]. Another interesting fact about $PtSe_2$ is that it can be grown under relatively low temperatures when compared to other TMDCs via vapour phase synthesis process. This increases the choice of substrates. It has been reported to be grown at temperatures as low as 100°C [7]. Although $PtSe_2$ emerged more than a century ago, its layer dependent properties and their respective applications are hardly explored. This chapter gives an outline of Raman spectroscopic studies of $PtSe_2$ of different thicknesses, both at room temperature and low temperature. It also deals with the task of using Raman spectroscopy to measure in-plane thermal conductivity. Theoretical calculation of electronic band structure and density of states of the monolayer, bilayer, tri layer, and bulk $PtSe_2$ has also been incorporated in this chapter.

The 1T phase of $PtSe_2$ belongs to space group $P\bar{3}m1$ and this phase is more stable when compared to 1H phase [8]. The Bravais lattice of 1T phase monolayer $PtSe_2$ is hexagonal and has D_{3d} point group symmetry. The monolayer is made up of three atomic sub layers, Pt layer is being sandwiched between two Se layers. The reported lattice vectors are $\vec{a} = 0.5a(\sqrt{3}x - y)$, $\vec{b} = 0.5a(\sqrt{3}x + y)$ and $\vec{c} = cz$ [9]. **Figure 1** shows the structure of $PtSe_2$, i.e. top view (a) and side view (b). The calculated lattice constant of the primitive unit cell of monolayer $PtSe_2$ is 3.70 Å which matches with the measured value of the same by STEM (scanning tunneling electron microscope) measurement [8, 10]. The primitive unit cell contains three atoms. The Pt-Se bond length for 1T phase is 2.52 Å and the distance between the top and bottom Se sub layers is 2.68 Å [10].

In monolayer 1T-$PtSe_2$ phase the bonding is completely covalent in nature with no net transfer of charge between the bonded atoms. The work function of single layer $PtSe_2$ is 5.36 eV and the values for other dichalcogenides like $MoSe_2$ and WSe_2 are 4.57 and 4.21 eV [8, 11]. The monolayer behaves as a semiconductor whereas the bulk behaves as a semi-metal [12, 13]. The phonon spectrum of mono layer 1T-$PtSe_2$ consists of nine phonon modes out of which six are optical and three are acoustic. **Figure 2a** shows the phonon modes in 1L1 T-$PtSe_2$. The six optical modes can be decomposed into - $\Gamma = 2E_g + 2E_u + A_{1g} + A_{2u}$. The in-plane modes (169 cm^{-1} for E_g and 218 cm^{-1} for E_u) are doubly degenerate and out of plane modes (200 cm^{-1} for A_{1g} and 223 cm^{-1} for A_{2u}) are singly degenerate, the ($A_{1g} + E_g$) modes are Raman active, $2A_u$ is infrared active and $2E_u$ is both infrared and Raman active. Out of these four modes, only two of them (E_g mode at 169 cm^{-1} and A_{1g} at 200 cm^{-1}) are prominent. **Figure 2b** shows their calculated relative intensities along with the modes of vibration [8]. The strong covalent

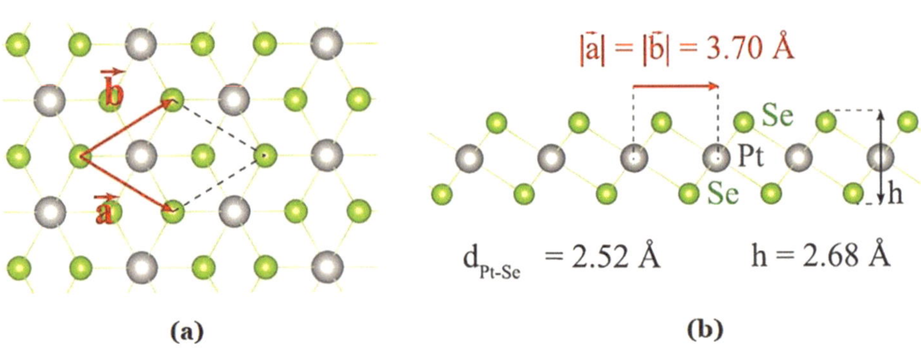

Figure 1.
(a) Top view and (b) side view of 1T phase $PtSe_2$. Grey atoms represent Pt and green atoms represent Se, respectively [8].

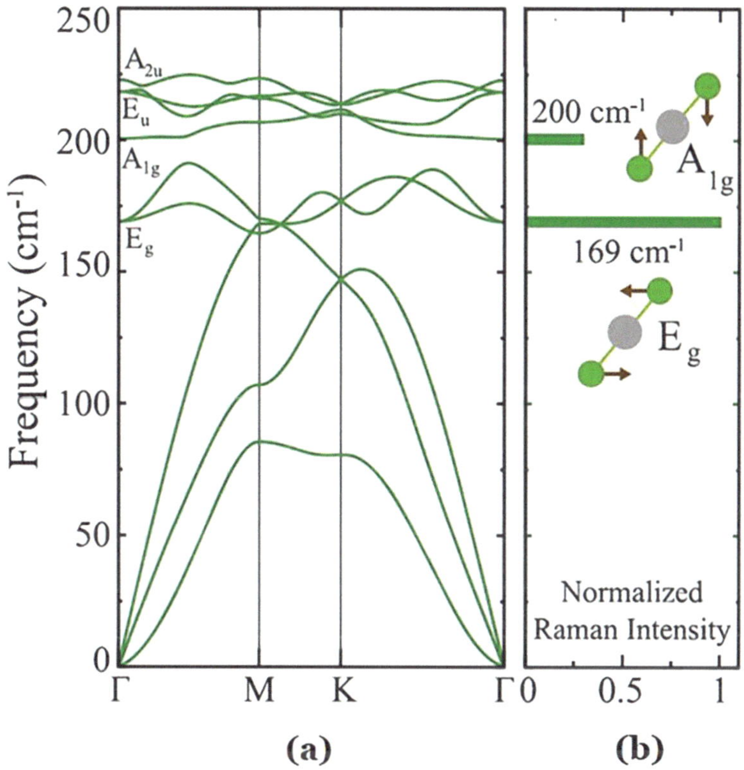

Figure 2.
(a) The phonon band diagram and (b) normalized Raman intensity of 1-layer 1T-PtSe$_2$ [8]. (Grey atoms – Pt and green atom – Se).

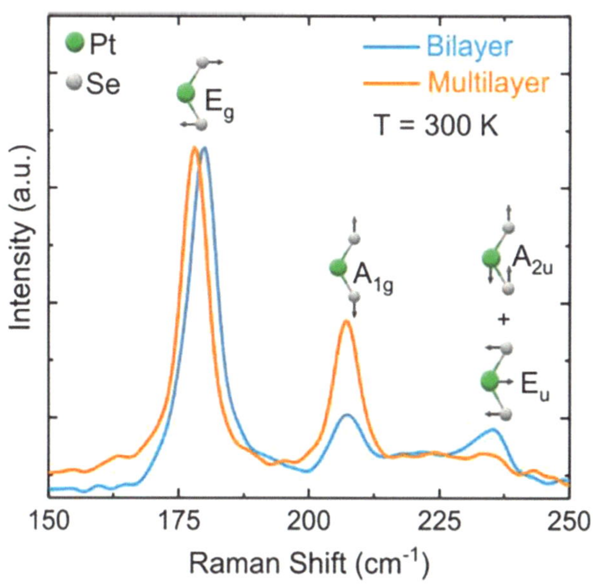

Figure 3.
Raman spectrum of bilayer and multilayer (~5 nm thick) CVD grown PtSe$_2$ at room temperature measured using laser wavelength of 532 nm [14].

nature of the Pt-Se bond makes the in plane vibrational modes more intense when compared to out of plane modes.

Figure 3 shows the Raman spectra of both bilayer and multilayer (~5 nm thick) CVD grown $PtSe_2$ [14]. In bilayer $PtSe_2$, two prominent peaks are observed which are centred at 179 and 207 cm^{-1}. These peaks are associated with first order phonon emission of in-plane and out of plane vibrational modes i.e. E_g and A_{1g} modes. Additionally, another less prominent peak is recorded at 235 cm^{-1}. This low intensity peak arises due to the longitudinal optical mode and can be separated into two vibrations. These two vibrations correspond to first order two phonon emission for out of plane (A_{2u}) and in plane (E_u) vibrations of Pt and Se atoms [14]. O'Brien et al. [15], reported the above mentioned peaks to be centred at 175 cm^{-1} (E_g), 205 cm^{-1} (A_{1g}), and 230 cm^{-1} (LO), and they also showed that these modes have approximately constant peak position for laser excitation of 488, 532, and 633 nm. The differences between theoretically derived (previously discussed) phonon modes and experimentally recorded ones can be attributed to the fact that $PtSe_2$ layers are not pristine with definite thickness [15].

2. Density functional theory study of layered $PtSe_2$

We employed density functional theory (DFT) to calculate the electronic band structure and density of states of mono layer, bi-layer, tri-layer, and bulk 1T -phase $PtSe_2$. This section deals with the theoretical calculations. DFT is a quantum modeling method used to investigate the properties of chemical systems, including atoms and molecules (i.e., many body systems), particularly the electronic structure properties. It is an ab inito method for solving the Schrodinger equations for many-electron systems which are defined by the electron density. The approach taken is, instead of using a many-body wave function, one-body density is used as the fundamental variable. Since the electron density $n(r)$ is a function of only three spatial coordinates, rather than 3N coordinates of the wave function, DFT is computationally feasible for small to large systems. The root of DFT comes from two theorems given by Hohenberg and Kohn who considered interacting electrons in an external field [16, 17]. The theorems state that the ground state energy is a unique function of electron density, allowing us to work in three dimensions than in 3N dimensions. Only the electron density that minimizes the energy of functional is taken, with the assumption that the function is known. HK theorems sound simple but can be applied only under certain circumstances but they are not considered in practice as densities of atoms do not obey these constraints. Taking into consideration some of the fundamental issues, Kohn and Sham reduce the problem to noninteracting electrons moving in an effective potential, leading to a set of self-consistent, single particle equations known as Kohn-Sham equations that contains exchange correlation potential. The result of DFT calculations depends on the choice of exchange-correlation functional. In terms of increasing accuracy, we have LDA, GGA, meta-GGA, and hybrid functionals. Some functionals give good results for one system and some for another. More accurate functionals consume more computational resources with a trade-off between accuracy and speed of calculation.

2.1 Computational methods

All the computational calculations were performed using the DFT with the projector augmented wave (PAW) pseudopotentials available with quantum-espresso [18]. Which is an integrated suite of open-source codes for the electronic structure calculation and materials modeling at the nanoscale. We used the generalized gradient approximation (GGA) of the Perdew-Burke-Ernzerhof (PBE) functional for the calculation of exchange and correlation potential. The van der Waals correction for the layered structures was taken into account using the DFT-D_2 method as proposed by Grimme [19]. The arrangement of layers was taken such that it has the lowest ground state energy value and the top-to-top (AA) stacking order is the most favorable one with the interlayer distance calculated to be 2.44 Å. The kinetic energy cut-off for a plane-wave basis set was taken to be 800 eV.

The convergence criteria for self-consistent calculations for electronic structures were set to 10^{-6} eV. For the optimized geometrical configurations, the energy convergence criterion was set to 10^{-5} eV, structure relaxations were conducted until the residual force acting on each atom is less than 0.01 eV/Å and pressure values less than 1 kbar. The sampling of the first Brillouin zone was done using Γ-centered k-point mesh of 15 × 15 × 15 and 15 × 15 × 1 for bulk and thin film structures, respectively. A region of at least 14 Å vacuum space was added in the z-direction to minimize the interaction between the neighboring atomic layers.

2.2 Structure and electronic properties

The bulk phase of $PtSe_2$ has 1T phase with tetragonal symmetry having space group P-3 m1 and the lattice constant, after optimization, found to be a = 3.77 Å, c = 5.52 Å, a significant close to the experimental value and similar other reported values as discussed earlier. The monolayer structure is composed of three atomic sublayers with a Pt layer sandwiched between two Se layers. The lattice constant is calculated to be 3.70 Å which is in agreement with previously reported data, and the vertical distance between the upper and lowermost layer of Se is about 2.65 Å close to the reported value of 2.53 Å [12]. While moving from bilayer to bulk structure, the interlayer distance also decreases, along with the trend of increasing layer-layer interaction which enlarges the covalent Pt-Se bond. The lattice constant of monolayer, bilayer and trilayer is found to be 3.70 Å, 3.73 Å and 3.74 Å respectively and for bulk it is 3.78 Å.

Figure 4 shows the electronic band structure of 1L,2L, 3L and bulk $PtSe_2$.The electronic structure calculations using the PBE functional show the transition from semiconductor to (semi)metal behavior of the material. The calculated bandgap of monolayer is around 1.39 eV, close to the experimental value [20]. While moving from monolayer to bilayer, the electronic band gap rapidly decreases to 0.38 eV. It is also found that $PtSe_2$ crystals having a thickness larger than two layers exhibit metallic behavior. Looking at the band structures, one can see that 1T-$PtSe_2$ monolayer has its valence band maximum (VBM) at Γ point and conduction band minimum (CBM) within Γ-M point. While going from monolayer to bilayer and higher layers, we observe that position of CBM at Γ-M point is fixed and VBM shifted from Γ point to within K-Γ high symmetry point. The reason behind this shift might be due to the non-periodicity of layered structure along the growth direction which is different from the band structure of bulk 3D structure of same material, as the energy band structure is strongly dependent on the crystal periodicity. The decrease in band energy of CB states and increase in VB states leads to metallization starting from trilayer [21, 22].

The DOS variation with different layers is shown in **Figure 5a–d**. In the DOS plots, for monolayer (**Figure 5a**), peak appears for the VBM, depicting a greater number of bands near the VBM, and flat near 0 eV. For the bilayer (**Figure 5b**), the

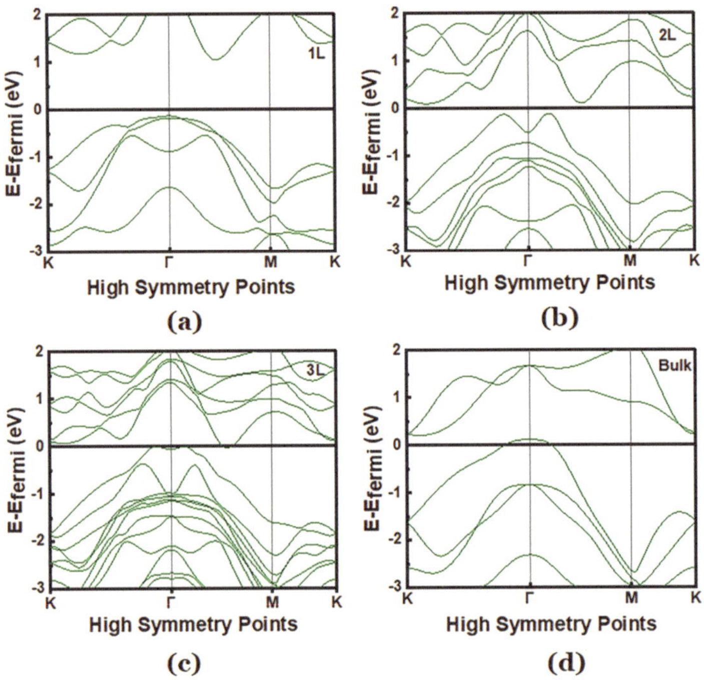

Figure 4.
Calculated electronic band structure of (a) 1L, (b) 2L, (c) 3L and (d) bulk $PtSe_2$.

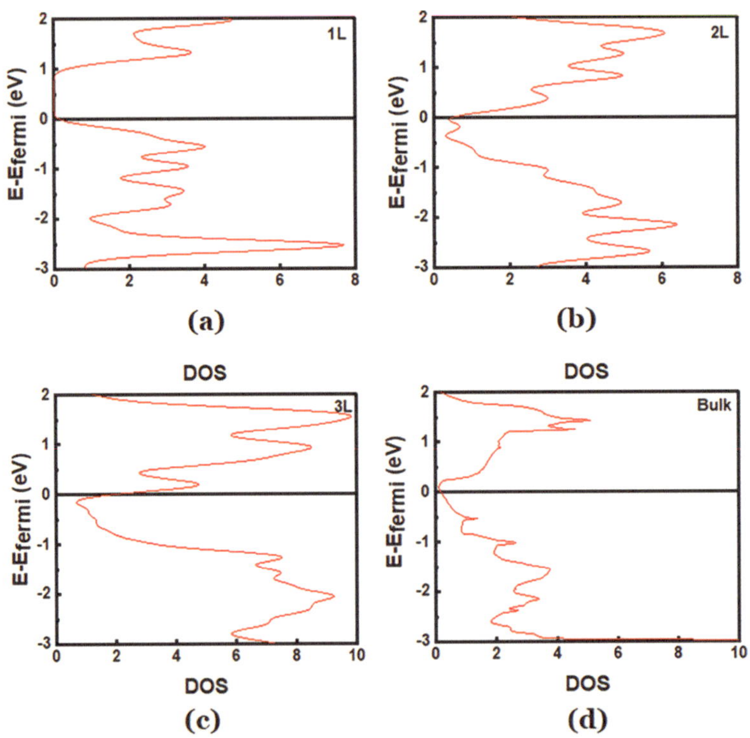

Figure 5.
Calculated density of state of (a) 1L, (b) 2L, (c) 3L and (d) bulk $PtSe_2$.

peak disappears near the VBM, where the band involves two peaks near Γ point. This remains until peak appears around fermi energy when it comes to more than three layers. It is also noticed that DOS under VBM is small for multilayers due to the large splitting between the first valence band with the second valence band [23, 24]. In this study, we were able to observe an increase in band gap with the decrease in layer numbers from bulk down to monolayer structures. Unlike other TMDs like MX_2 (M = Mo and W; X = S and Se) which are direct bandgap semiconductors at monolayer, there is no shift from indirect-to-direct band gap with decrease in number layers from bulk to monolayer limit. This may be due to the difference in crystal structure i.e. – MX_2 has 2H structure and $PtSe_2$ has 1T structure. We were able to observe the inverse relationship between the band gap and number of layers, which is governed by factors such as quantum confinement effect and interlayer interaction.

3. In plane thermal conductivity of layered $PtSe_2$

The thermal conductivity of layered materials can be measured by employing Raman spectroscopy which is non-destructive in nature. It is fairly common to employ this method to measure the in-plane thermal conductivity in many 2D materials like graphene, *h*-BN and other TMDCs [25–28]. There are several advantages for this method, such as measurement can be done for 2D material of different thicknesses over any substrates and also on suspended ones. In addition, the effect of substrate on the thermal conductivity can also be studied. Here, we employed Raman spectroscopy to study the thermal conductivity of multilayer mechanically exfoliated $PtSe_2$. The parent crystal used for mechanical exfoliation was sourced from 2D semiconductors, USA. Exfoliation was done over SiO_2/Si substrate with SiO_2 layer being 290 nm thick. A Horiba HR800 UV Raman spectrometer was used to acquire all Raman spectra which had 1800 lines/mm grating and 100X objective (0.9 NA) with a spot size of ~1 μm. The excitation wavelength was 488 nm f0r all the recorded Raman spectra. For low temperature Raman spectroscopy, Linkam liquid nitrogen cooling stage was used. During low temperature measurement, a long working distance 50× objective with 0.5 NA was used. The Raman plots are fitted with the Lorentzian function where the solid lines are fitted data and dots are raw data. We measured the thickness of the flakes by using atomic force microscopy (AFM), which was done by using a Park systems NX10 model. The AFM images were recorded in non-contact mode. Tip radius was less than 10 nm having a force constant of 42 N/m and frequency of 330 kHz. The height has been measured at 10 places to estimate the standard deviation in the measurement. The optical micrographs were taken by using a Nikon Eclipse LV100ND microscope.

We used three differently thick multilayer flakes (named as flake 1, flake 2 and flake 3) for this study. The Raman spectra of flake 1 at different laser power and at different temperatures is depicted in **Figure 6a** and **b** respectively. The laser power was varied from 0.227 to 1.08 mW whereas the temperature was varied from 107 to 293 K. Both the in-plane (E_g) and out of plane (A_{1g}) modes shifted to lower wave numbers with the rise in temperature. This shift is due to the thermally bond softening [29]. With higher power both the modes shifted to higher wavelength. **Figure 7a** and **b** shows the AFM and optical micrograph of flake 1 respectively. Although the optical image shows a uniform contrast over the whole flake, it is clear from the AFM image that the edges have different thicknesses. The inset of the AFM image shows

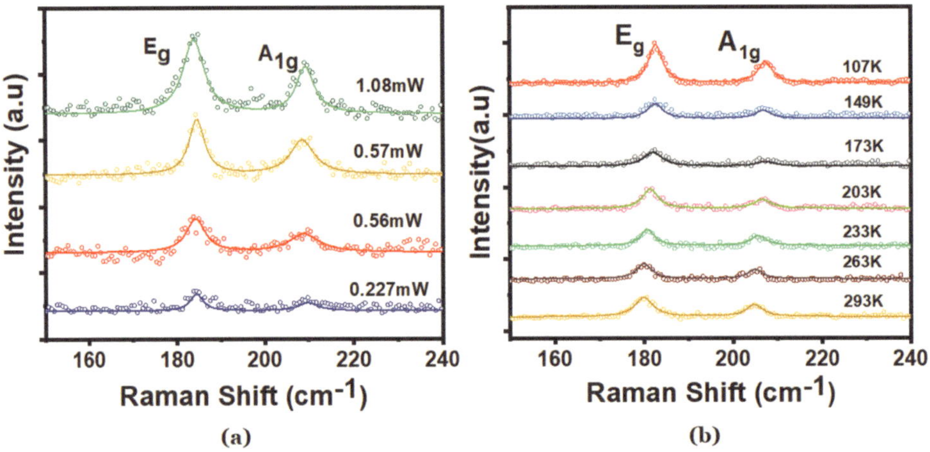

Figure 6.
Power dependent Raman spectra (a) and temperature dependent Raman spectra (b) of flake 1 measured at fixed power of 2.27 mW.

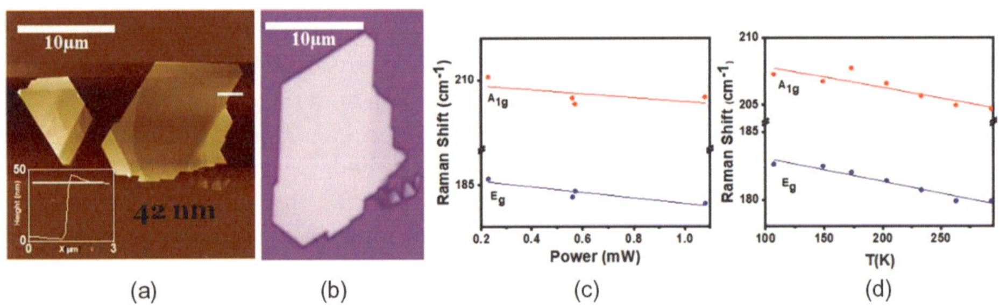

Figure 7.
(a) AFM image of the flake 1, inset showing the height profile, (b) Optical micrograph of the same flake. (c) and (d) Variation of the in plane (E_g) and out of plane (A_{1g}) with power and temperature respectively.

the height profile, the thickness is about 42 nm ± 2 nm (at the middle part where the thickness is uniform). Which corresponds to approximately 54 layers, each layer being 0.8 nm thick [10]. The Raman spectra in **Figure 6** don't show the LO mode, as the flake is about 42 nm in thickness whereas the LO mode is suppressed due to the bulk nature of the flake.

The variation of the E_g and A_{1g} (for flake 1) with laser power and temperature is shown in **Figure 7c** and **d** respectively. Both the plot is linearly fitted and have a negative slope. The range of power was chosen such that the excitation laser doesn't damage the flake [30, 31]. The negative slope in power dependent Raman spectra for both E_g and A_{1g} implies that the peaks shift to lower wave numbers. This is due to the fact that with increase in power local heating increases and the Pt-Se bonds softens. The thermal conductivity can be deduced by using these two plots (**Figure 7a** and **d**). The power (P) dependent peak position (ω) is linear in the low power range and is governed by [32]:

$$\Delta\omega = \omega(P_2) - \omega(P_1) = \chi(P_2 - P_1) = \chi_P \, \Delta P \tag{1}$$

So, the power coefficient is given by $\chi_P = \dfrac{\Delta\omega}{\Delta P}$.

The power coefficient of both the E_g and A_{1g} modes have been calculated through a linear fit of power dependent peak shift of these modes (as depicted in **Figure 7c**). The temperature dependence of both the in-plane (E_g) and out of plane (A_{1g}) Raman modes can be stated as [32, 33]:

$$\omega(T) = \omega_0 + \alpha_1 T + \alpha_2 T^2 \quad (2)$$

Where ω_0 is the frequency of E_g and A_{1g} at T = 0 K, α_1 and α_2 are the first and second order temperature coefficients. The second order coefficient is applicable at high temperatures, so for the present case, this can be neglected. After neglecting the second order coefficient, the equation was used to linearly fit the evolution Raman peak position with temperature in **Figure 7b**. The first order temperature and power coefficient can be used to get the thermal conductivity (K) arising from a particular mode of vibration. For 2D materials the expression used by using α_1 and χ_P and is given by [34]:

$$K = \alpha_1 \chi_P^{-1} \left(\frac{1}{2\pi h} \right) \quad (3)$$

where h is the thickness of the flake.

The slope of linear fit corresponding to E_g mode in **Figure 7c** and **d** i.e. χ_P and α_1 is −1.654 cm^{-1}/mW and −0.017 cm^{-1}/K respectively. Applying these values to Eq. (3) we get in-plane thermal conductivity for flake 1, which is –38.90 W/mK ± 17.43 W/mK. The error in thermal conductivity was calculated by taking into consideration the error of the slopes in **Figure 7c** and **d** and also for the thickness of flake 1. The range of error is consistent with the current literature for this method [35].

We investigated another two flakes named flake 2 and flake 3. The Raman spectra of flake 2 at different laser power and at different temperature is shown in **Figure 8a** and **b** respectively. **Figure 8c** shows the AFM image, with the optical micrograph and height profile in the inset. The thickness of flake 2 as derived from the height profile is 59 nm ±2 nm. Here too the absence of the LO mode is due to the bulk nature of the flake. The evolution of the E_g and A_{1g} Raman modes with power and temperature is depicted in **Figure 9a** and **b** respectively.

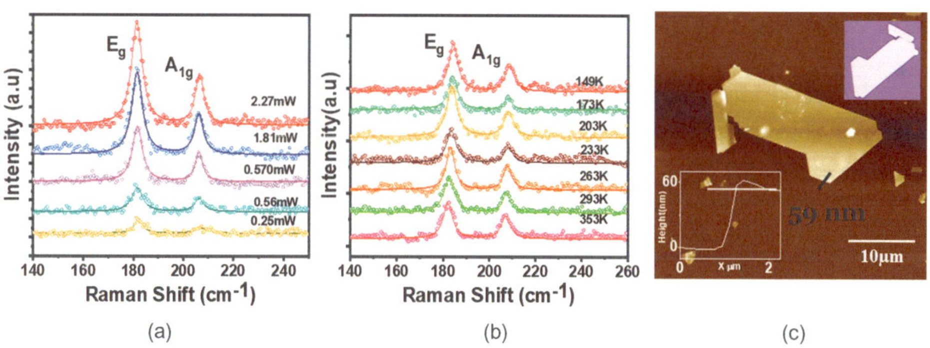

Figure 8.
(a) Raman spectra at different laser power, (b) at different temperature with fixed laser power of 2.27 mW. (c) AFM image, inset shows the optical micrograph and the height profile of flake 2.

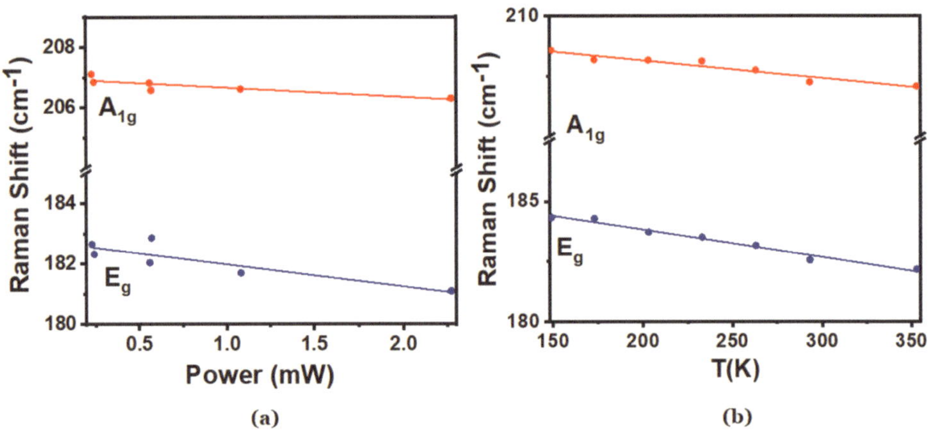

Figure 9.
Dependence of Raman peak shift (both E_g and A_{1g}) with (a) power and (b) with temperature of flake 2.

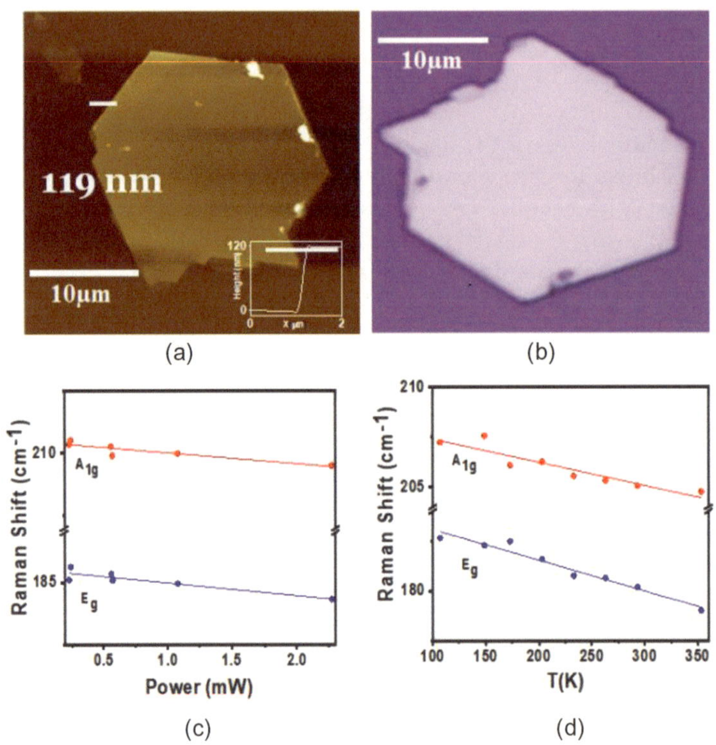

Figure 10.
(a) and (b) AFM and optical micrograph of flake 3, inset showing the height profile. (c) and (d) Variation of the Raman peak shift (both E_g and A_{1g}) with laser power and temperature respectively.

The plots are linearly fitted as discussed in previous section. Employing Eq. (3) we find the thermal conductivity of flake 2 is 40.42 W/mK ± 16.37 W/mK. **Figure 10a** and **b** shows the AFM, optical micrographs and evolution of the E_g and A_{1g} modes with power and temperature (**Figure 10c** and **d** respectively) of flake 3. The thickness of flake 3 is 119 nm ± 3 nm. Using α_1 and χ_P from Eq. (3) we get the thermal conductivity, which is 39.30 W/mK ± 14.96 W/mK.

Flake	h (nm)	α_1 (cm^{-1}/K)	χ_P (cm^{-1}/mW)	K (W/mK)
1	42	−0.017	−1.654	38.90
2	59	−0.011	−0.734	40.42
3	119	−0.015	−0.511	39.30

Table 1.
Comparison of thermal conductivity, α_1 and χ_P of the three flakes for E_g mode.

Table 1 shows the comparison between the thermal conductivity, α_1 and χ_P of the three flakes. The thermal conductivity of all the flakes is approximately similar. This shows that $PtSe_2$ over 40 nm thick has almost constant thermal conductivity. The first order temperature coefficient is also fairly constant for all three flakes. The calculated thermal conductivity at 300 K for monolayer $PtSe_2$ is about 18 W/mK [36]. The saturation value for thicker $PtSe_2$ over SiO_2/Si substrates can be approximately taken as 40 W/mK.

4. Effect of laser polarization over different thicknesses of $PtSe_2$

The state of laser polarisation during Raman spectroscopy affects the two prominent modes of vibration i.e., E_g (in plane) and A_{1g} (out of plane) mode. **Figure 11a** and **b** shows the 1 nm (a) and 5 nm CVD grown $PtSe_2$ under vertical (Z(YY)-Z) and horizontal (Z(YX)-Z) polarization [15]. For the 1 nm film, the intensity out of plane (A_{1g}) mode significantly varies under vertical and horizontal polarization, whereas the 5 nm film has considerably less variation. When the thickness is increased to 59 nm (flake 2, **Figure 11c**) the intensity is almost similar under both states of polarization. This is because for the thicker film, the contribution from the bottom layers is more prominent and makes the intensities similar. The state of polarization has no effect over the intensity of the in-plane mode for 1 nm and 59 nm, this is because of it's planar nature of vibration.

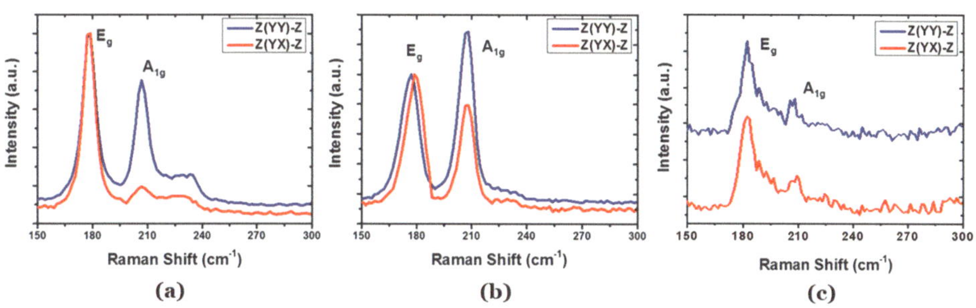

Figure 11.
Raman spectra of CVD grown (a) 1 nm and (b) 5 nm films of $PtSe_2$ [15] (c) flake 2 of present work under parallel and horizontal polarization.

5. Conclusion

2D $PtSe_2$ is an excellent material, its unique layer dependent properties. The fact that it is stable under ambient conditions unlike most other TMDCs makes it a great choice for scientific study. This book chapter gives an overview of the layer dependent properties of $PtSe_2$ like band gap, density of states and thermal conductivity. DFT was employed to study the electronic band structure of 1L, 2L, 3L and bulk $PtSe_2$, which showed that there is a drastic reduction of band gap when moving from monolayer to bilayer.

Optothermal method by using Raman spectroscopy was employed to explore the thermal conductivity of $PtSe_2$ flakes. The Raman study was carried out by both varying the power and temperature of the sample. The incident laser power was varied from 0.25 to 2.27 mW and the temperature of the sample was varied from 107 to 353 K, the power coefficient (χ_p) and the temperature coefficient (α_1) was calculated from these data. The thermal conductivity was obtained by using both coefficients for a specific thickness of the flake. The optothermal study reveals that the saturation thermal conductivity of $PtSe_2$ with thickness more than 40 nm is about 39–41 W/mK. Perpendicular and parallel polarization study was done for 59 nm thick flake, which reveals that both the in plane and out of plane modes didn't suffer any change in intensity in contrast with thinner flake (5 nm thick). The authors hope this book chapter shall aid the exploration endeavour regarding $PtSe_2$.

Acknowledgements

PKN acknowledges the financial support from the Department of Science and Technology, Government of India (DST-GoI), with sanction Order No. SB/S2/RJN-043/2017 under Ramanujan Fellowship. This work was also partially supported by Indian Institute of Technology Madras to the 2D Materials Research and Innovation Group and Micro-Nano and Bio-Fluidics Group under the funding for Institutions of Eminence scheme of Ministry of Education, GoI [Sanction. No: 11/9/2019-U.3 (A)]

Author details

Nilanjan Basu[1,2], Vishal K. Pathak[1,3], Laxman Gilua[1,2] and Pramoda K. Nayak[1,2,3]*

1 Department of Physics, Indian Institute of Technology Madras, Chennai, India

2 2D Materials Research and Innovation Group, Indian Institute of Technology Madras, Chennai, India

3 Micro Nano and Bio-Fluidics Group, Indian Institute of Technology Madras, Chennai, India

*Address all correspondence to: pnayak@iitm.ac.in

References

[1] Novoselov KS, Geim AK, Morozov SV, Jiang D, Zhang Y, Dubonos SV, et al. Electric field effect in atomically thin carbon films. Science. 2004;**306**: 666-670. DOI: 10.1126/science.1102896

[2] Boehm HP, Setton R, Stumpp E. Nomenclature and terminology of graphite intercalation compounds. Carbon. 1986;**24**(2):241-245. DOI: 10.1016/0008-6223(86)90126-0

[3] Zeng L et al. Ultrafast and sensitive photodetector based on a $PtSe_2$/silicon nanowire array heterojunction with a multiband spectral response from 200 to 1550 nm. NPG Asia Materials. 2018;**10**(4):352-362. DOI: 10.1038/s41427-018-0035-4

[4] Yim C et al. High-performance hybrid electronic devices from layered $PtSe_2$ films grown at low temperature. ACS Nano. 2016;**10**(10):9550-9558. DOI: 10.1021/acsnano.6b04898

[5] Yim C et al. Wide spectral photoresponse of layered platinum diselenide-based photodiodes. Nano Letters. 2018;**18**(3):1794-1800. DOI: 10.1021/acs.nanolett.7b05000

[6] Boland CS et al. $PtSe_2$ grown directly on polymer foil for use as a robust piezoresistive sensor. 2D Materials. 2019;**6**(4):045029. DOI: 10.1088/2053-1583/ab33a1

[7] Su TY et al. Phase-engineered $PtSe_2$-layered films by a plasma-assisted selenization process toward All $PtSe_2$-based field effect transistor to highly sensitive, flexible, and wide-spectrum photoresponse photodetectors. Small. 2018;**14**(19):1-10. DOI: 10.1002/smll.201800032

[8] Kandemir A et al. Structural, electronic and phononic properties of $PtSe_2$: From monolayer to bulk. Semiconductor Science and Technology. 2018;**33**(8):085002. DOI: 10.1088/1361-6641/aacba2

[9] Ryu GH, Chen J, Wen Y, Warner JH. In-situ atomic-scale dynamics of thermally driven phase transition of 2D few-layered 1T $PtSe_2$ into ultrathin 2D nonlayered PtSe crystals. Chemistry of Materials. 2019;**31**(23):9895-9903. DOI: 10.1021/acs.chemmater.9b04274

[10] Zhou J et al. Epitaxial synthesis of monolayer $PtSe_2$ single crystal on $MoSe_2$ with strong interlayer coupling. ACS Nano. 2019;**13**(10):10929-10938. DOI: 10.1021/acsnano.8b09479

[11] Gong C, Zhang H, Wang W, Colombo L, Wallace RM, Cho K. Band alignment of two-dimensional transition metal dichalcogenides: Application in tunnel field effect transistors. Applied Physics Letters. 2013;**103**(5):053513. DOI: 10.1063/1.4817409

[12] Wang Y et al. Monolayer $PtSe_2$, a new semiconducting transition-metal-dichalcogenide, epitaxially grown by direct selenization of Pt. Nano Letters. 2015;**15**(6):4013-4018. DOI: 10.1021/acs.nanolett.5b00964

[13] Buchkov K et al. Synthesis and characterization of 2D platinum diselenide. Journal of Physics: Conference Series. 2020;**1492**(1):012022. DOI: 10.1088/1742-6596/1492/1/012022

[14] Gulo DP, Yeh H, Chang WH, Liu HL. Temperature-dependent optical and vibrational properties of $PtSe_2$ thin films. Scientific Reports. 2020;**10**(1):1-12. DOI: 10.1038/s41598-020-76036-y

[15] O'Brien M et al. Raman characterization of platinum diselenide

thin films. 2D Materials. 2016;**3**(2):021004. DOI: 10.1088/2053-1583/3/2/021004

[16] Chu CH, Leung CW. The convolution equation of Choquet and Deny on [IN]-groups. Integral Equations and Operator Theory. 2001;**40**(4):391-402. DOI: 10.1007/BF01198136

[17] Atoms DT, Parr MRG, Yang W. Density-Functional Theory of Atoms and Molecules. Vol. 471989. New York: Oxford University Press. p. 10101

[18] Giannozzi P et al. QUANTUM ESPRESSO: A modular and open-source software project for quantum simulations of materials. Journal of Physics: Condensed Matter. 2009;**21**(39):395502. DOI: 10.1088/0953-8984/21/39/395502

[19] Grimme S. Semiempirical GGA-type density functional constructed with a long-range dispersion correction. Journal of Computational Chemistry. 2006;**27**(15):1787-1799. DOI: 10.1002/jcc.20495

[20] Zhao X, Huang R, Wang T, Dai X, Wei S, Ma Y. Steady semiconducting properties of monolayer $PtSe_2$ with non-metal atom and transition metal atom doping. Physical Chemistry Chemical Physics. 2020;**22**(10):5765-5773. DOI: 10.1039/c9cp06249a

[21] Shi J et al. Chemical vapor deposition grown large-scale atomically thin platinum diselenide with semimetal-semiconductor transition. ACS Nano. 2019;**13**(7):8442-8451. DOI: 10.1021/acsnano.9b04312

[22] Villaos RAB et al. Thickness dependent electronic properties of Pt dichalcogenides. npj 2D Materials and Applications. 2019;**3**(1):1-8. DOI: 10.1038/s41699-018-0085-z

[23] Hong Z. Tunable structure and electronic properties of multilayer $PtSe_2$. Journal of Physics: Conference Series. 2019;**1411**(1):012019. DOI: 10.1088/1742-6596/1411/1/012019

[24] Absor MAU, Santoso I, Harsojo K, Abraha FI, Saito M. Defect-induced large spin-orbit splitting in monolayer $PtSe_2$. Physical Review B. 2017;**96**(11):1-6. DOI: 10.1103/PhysRevB.96.115128

[25] Zhou H et al. High thermal conductivity of suspended few-layer hexagonal boron nitride sheets. Nano Research. 2014;**7**(8):1232-1240. DOI: 10.1007/s12274-014-0486-z

[26] Malekpour H, Balandin AA. Raman-based technique for measuring thermal conductivity of graphene and related materials. Journal of Raman Specroscopy. 2018;**49**(1):106-120. DOI: 10.1002/jrs.5230

[27] Peimyoo N, Shang J, Yang W, Wang Y, Cong C, Yu T. Thermal conductivity determination of suspended mono- and bilayer WS2 by Raman spectroscopy. Nano Research. 2015;**8**(4):1210-1221. DOI: 10.1007/s12274-014-0602-0

[28] Yan R et al. Thermal conductivity of monolayer molybdenum disulfide obtained from temperature-dependent Raman spectroscopy. ACS Nano. 2014;**8**(1):986-993. DOI: 10.1021/nn405826k

[29] Yin S et al. Thermal conductivity of few-layer PtS_2 and $PtSe_2$ obtained from optothermal Raman spectroscopy. Journal of Physical Chemistry C. 2021;**125**(29):16129-16135. DOI: 10.1021/acs.jpcc.1c02522

[30] Laikhtman A, Hoffman A. Laser power effects on the Raman spectrum of isolated diamond chemical vapor deposition particles. Journal of Applied

Physics. 1997;**82**(1):243-248. DOI: 10.1063/1.365804

[31] Sahoo S, Gaur APS, Ahmadi M, Guinel MJF, Katiyar RS. Temperature-dependent Raman studies and thermal conductivity of few-layer MoS2. Journal of Physical Chemistry C. 2013;**117**(17): 9042-9047. DOI: 10.1021/jp402509w

[32] Singh MP, Ryntathiang S, Krishnan S, Nayak PK. Study of thermal conductivity in two-dimensional Bi_2Te_3 from micro-Raman spectroscopy. Current Chinese Science. 2021;**1**(4):453-459. DOI: 10.2174/2210298101666210412101 104

[33] Park D, Park S, Jeong K, Jeong HS, Song JY, Cho MH. Thermal and electrical conduction of single-crystal Bi_2Te_3 nanostructures grown using a one step process. Scientific Reports. 2016;**6**(January):13-15. DOI: 10.1038/ srep19132

[34] Balandin D, Teweldebrhan AA, Lau CN, Ghosh S, Bao W, Calizo I, et al. Superior thermal conductivity of single-layer graphene. Nano Letters. 2008;**8**(3):902-907. DOI: 10.1021/ nl0731872

[35] Chen S et al. Raman measurements of thermal transport in suspended monolayer graphene of variable sizes in vacuum and gaseous environments. ACS Nano. 2011;**5**(1):321-328. DOI: 10.1021/ nn102915x

[36] Guo SD. Biaxial strain tuned thermoelectric properties in monolayer $PtSe_2$. Journal of Materials Chemistry C. 2016;**4**(39):9366-9374. DOI: 10.1039/ c6tc03074b

Chapter 4

Functional Mimics of Glutathione Peroxidase: Spirochalcogenuranes, Mechanism and Its Antioxidant Activity

Devappa S. Lamani

Abstract

The present chapter describe a series of synthetic organoselenium compounds such as ebselen analogues, diaryl selenides, spirodioxyselenurane, spirodiazaselenuranes and its Glutathione peroxidise (GPx) catalytic activity. These ebselen related compounds either by modifying the basic structure of ebselen or incorporating some structural features of the native enzyme, a number of small-molecules of selenium compounds as functional mimics of GPx are discussed. In addition to this, spirodioxyselenuranes and spirodiazaselenuranes are important class of hypervalent selenium compounds, whose stability highly depends on the nature of the substituents attached to the nitrogen atom. The glutathione peroxidase (GPx) mimetic activity of all the selenium compounds showed significantly by facilitating the oxidation of the selenium centre. In contrast to this, ebselen analogue shows significant antioxidant activity compared with spirodiazaselenuranes and its derivatives.

Keywords: spirodiazaselenuranes, antioxidant activity, selenoenzymes, ebselene

1. Introduction

Selenium has been discovered by the Swedish scientist Jons Jakob Berzelius in 1818. The chemistry of selenium, the next element to sulfur in the chalcogen group, is very less explored as compared to the chemistry of sulfur [1]. The diethyl selenide was synthesized by Lowig in 1836 and it was obtained in pure form in 1869 ass first synthetic selenium compound [2, 3]. Selenium chemistry was initially mainly focused on the synthesis of simple diselenides (RSeSeR), selenols (RSeH) etc. However, due to the unpleasant odor of selenols and aliphatic selenides, and also its toxicity the selenium chemistry faced a serious setback. Furthermore, due to toxicity of selenium was associated with diseases such as liverstock disease [4], intoxication in experimental animals [5–7] etc., therefore, it was considered a toxic element. In 1954 by Pinsent was established with the beneficial effect of selenium for living organisms the discovery that certain bacteria grew faster in selenium fortified medium [8]. However, the exact role of selenium responsible for the growth of bacteria was not clear. Almost after

20 years of this discovery, in 1973, it was found that two bacterial enzymes, formate dehydrogenase and glycine reductase contain selenium in their active sites [9, 10]. Flohe and co-workers was discovered almost at the same time the importance of selenium to mammals [11]. They found that the mammalian enzyme glutathione peroxidase (GPx), contains a selenocysteine residue in its active site. Nowadays the major selenoenzymes discovered to date include formate dehydrogenases [12], hydrogenases [13–16], glycine reductase [17] iodothyronine deiodinases (ID) [18–22], thioredoxin reductases (TrxR) [23–26], selenophosphate synthetase [27], and selenoprotein P [28, 29], glutathione peroxidase (GPx) [30–33].

1.1 Glutathione peroxidase

Glutathione peroxidase (GPx) an mammalian enzyme, contains selenocysteine residue in its active site, For the last four decades, an extensive research has been carried out on the mammalian antioxidant enzymes GPx [34]. The cGPx utilizes glutathione (GSH) as reducing substrate exclusively for the reduction of H_2O_2 and organic hydroperoxides such as *tert*-butyl hydroperoxide (t-BuOOH) and cumene hydroperoxide (Cum-OOH). This enzyme exhibits good activity with all phospholipid hydroperoxides, fatty acid hydroperoxides, t-BuOOH, Cum-OOH, cholesterol hydroperoxides, and H_2O_2. The crystal structure of GPx indicates that the Sec residue (Sec45) forms a 'catalytic triad' with other two amino acids, glutamine (Gln80) and tryptophan (Trp158) (**Figure 1**) [36].

The crystal structure of the seleninic acid form of human pGPx also indicates that Gln79 and Trp153 are located within hydrogen bonding distance of the selenium atom (**Figure 1**). These residues appear to play an important functional role in their catalytic mechanisms.

A catalytic cycle of GPx (**Figure 2**) starts with the oxidation of the selenol (ESeH) moiety of Sec residue by peroxide to generate the selenenic acid (ESeOH) [37, 38], which reacts with cellular thiol (glutathione, GSH) to generate a selenenyl sulfide intermediate (ESeSG).

Another equivalent of GSH cleaves the -Se-S- bond in the selenenyl sulfide intermediate to regenerate the selenol with elimination of glutathione disulfide (GSSG). The cellular level concentration of GSH is maintained by glutathione reductase (GR) [39], which reduces GSSG to GSH by using NADPH as cofactor. The overall catalytic mechanism, two equivalents of NADPH is consumed to reduce one equivalent of peroxide. At very high concentrations of hydroperoxide the selenium centre in GPx may

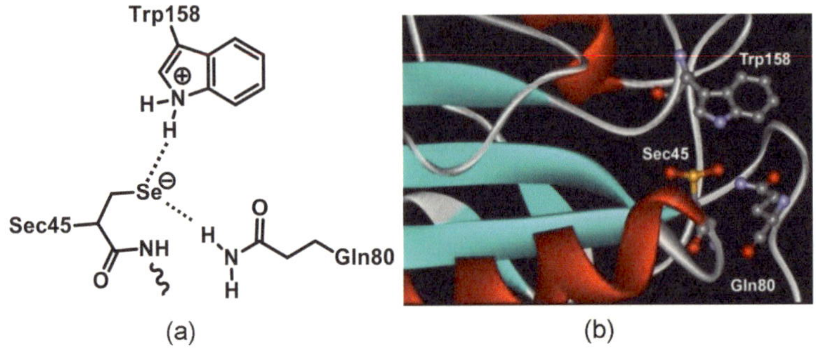

Figure 1.
(a) Catalytic triad at the active site of GPx; (b) active site of glutathione peroxidase (PDB code 1GP1) determined by X-ray crystallography [35].

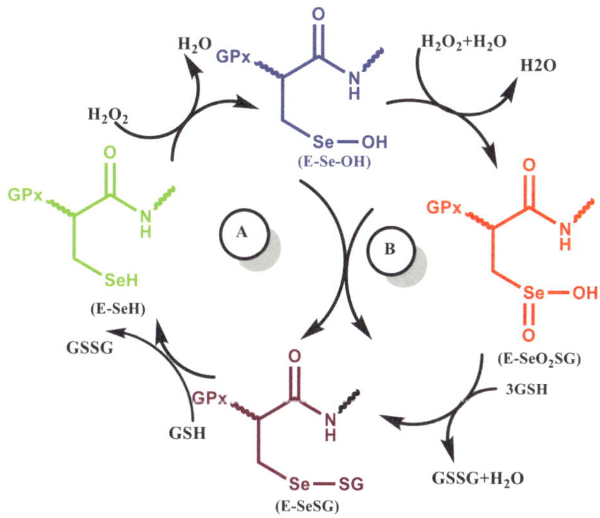

Figure 2.
Proposed mechanism for the GPx-catalyzed reduction of H_2O_2.

be overoxidized to produce seleninic acid ($ESeO_2H$) and selenonic acid ($ESeO_3H$). Whereas the oxidation of selenenic acid to seleninic acid is reversible in the presence of GSH, the further oxidation to selenonic acid may inactivate the enzyme.

2. Mimics and models of glutathione peroxidase

2.1 Ebselen analogues as GPx mimics/models

Synthetic selenium compounds with significant GPx activity have potential therapeutic applications (**Figure 3**). The first synthetic compound that has been shown to mimic the GPx activity was ebselen ([2-phenyl-1,2-benzisoselenazole-3)-(2*H*)-one] (**1**) [40–43]. Furthermore, the synthesis of such compounds may help in understanding the chemistry at the active site of GPx. The initial success of ebselen was mainly due to its very low toxicity and high stability of the selenazole moiety does

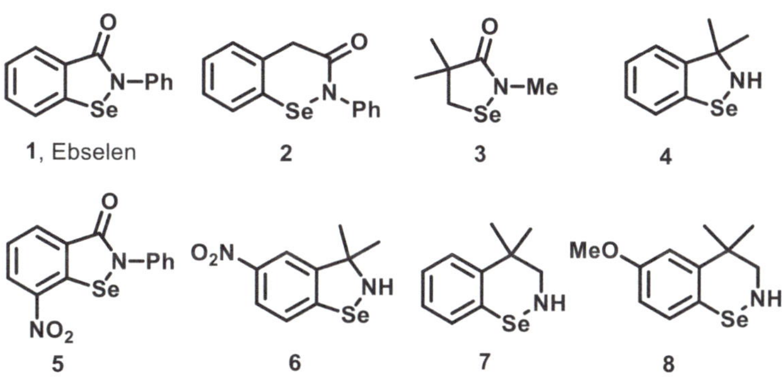

Figure 3.
Some representative examples of Ebselen analogues as GPx mimics.

not allow the elimination of selenium during the biotransformations. Therefore, the selenium metabolism of this compound does not interfere with the organism and as a result ebselen used in clinical trials for the treatment of patients suffering from active ischemia stroke.

In literature there were a several animal model studies have demonstrated that ebselen reduces oxidative stress in ischemia-reperfusion in heart and that it exhibits promising neuroprotective effect in brain. In addition to this, ebselen can be toxic to cells suggested in recent evidences. It has been shown that ebselen inhibits certain cell growth and induces apoptosis. However, the mechanism underlying the toxicity of ebselen is not known, the cellular glutathione (GSH) level appears to be depleted by ebselen. The GSH depletion increases the susceptibility of cells to oxidant injury as the reduced GSH is important for cell survival.

After the discovery that ebselen exhibits significant antioxidant activity by mimicking the active site of GPx, much attention has been devoted to the design and synthesis of novel analogues of ebselen. The ebselen homolog **2**, tetrahedral carbon is incorporated into the heterocycle, retains the Se-N bond essential for the GPx activity. The selenazole model system **3** has been used extensively to understand the anti-oxidant redox chemistry of selenocysteine at the active site of GPx and several such compounds has been synthesized and evaluated for its GPx activity (**Figure 4**) [44].

According to this mechanism the corresponding selenenyl sulfide is mainly the reaction of ebselen **1** with a thiol (RSH). The obtained intermediate compound is found to be unstable in the assay system, and therefore, undergoes a disproportion-ation reaction to generate the stable diselenide. Subsequent reaction of with peroxide produces the selenenic acid and seleninic acid. During this mechanism when RSH is depleted in the reaction mixture, the seleninic acid and interestingly, the selenenic acid having a free N-H moiety undergoes cyclization to regenerate ebselen1 [45].

Back and co-workers [46], reported the catalytic cycle of di(3-hydroxy-propyl) selenide **9** and acts an efficient catalyst for the reduction of t-BuOOH in the presence of BnSH. The compound **9** involves the formation of an unusual spirodioxyselenurane

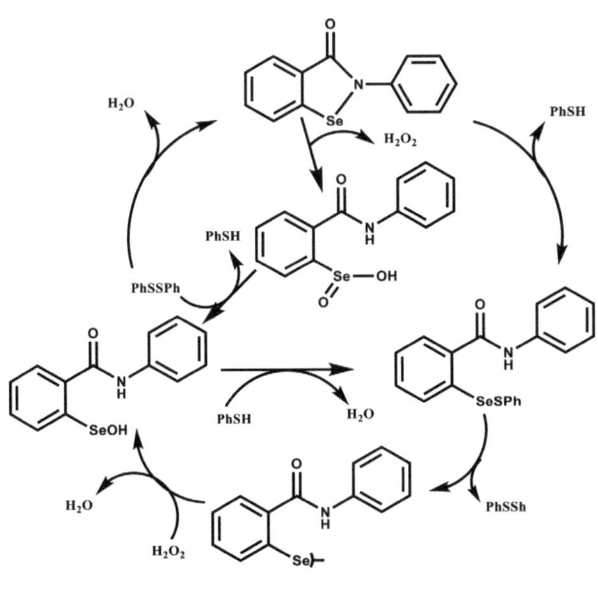

Figure 4.
Proposed catalytic cycle of ebselen and related compounds [45, 46].

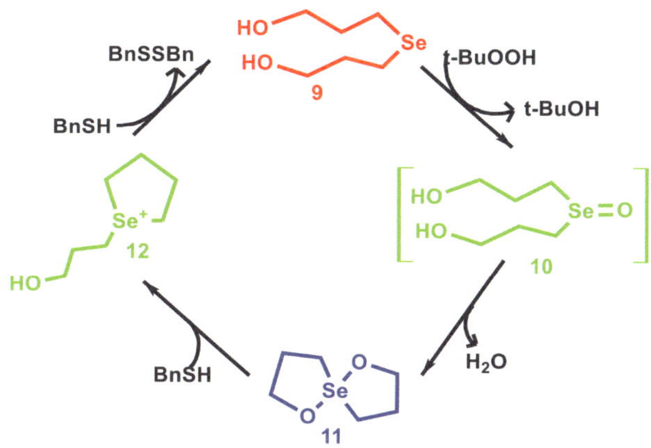

Figure 5.
Catalytic cycle of selenide 9 [47].

10. The oxidation of compound **9** with t-BuOOH produces the transient selenoxide **10**, which undergoes a spontaneous cyclization to produce the dioxyselenurane **11** isolated compound structure was confirmed by spectroscopic methods and single X-ray crystallography.

The reaction of **11** with BnSH produces an intermediate **12**, which upon reaction with second equivalent of BnSH regenerates the selenide **9** with elimination of BnSSBn (**Figure 5**). When t-BuOOH is present in the reaction mixture, compound **10** is recyclized to compound **11**. Although the reactivity of compound **9** was only about 15times higher than that of ebselen under indentical condition, the catalytic mechansium involves the formation of an unusual spiro compound [47].

3. Spirochalcogenuranes

3.1 Spirodioxyselenuranes/spirodiazaselenurane and its analogues as GPx mimics/models

Lesser and Weiss in 1914 reported the first example of a spirodioxyselenurane **13**. After this initial study, several spirodioxyselenuranes such as **14–19** have been reported in the literature [48–50]. This type of hypervalent selenium compounds attracted significant attention in recent years due to their interesting structural and stereochemical properties [51]. The selenium center in spirodioxyselenuranes generally shows trigonal bipyramidal geometry around central atom with the lone pair lying in the equatorial plane and the electronegative oxygen atoms occupying the apical positions [52]. In contrast to the well-studied spirodioxyselenuranes, spirodiazaselenuranes that contain two nitrogen substituents are extremely rare. Back and co-workers [53] few years ago, demonstrated the relative instability of spirodiazaselenuranes. They reported that the oxidation of the selenium center in 2,2′-selenobis(benzamide) by H_2O_2 does not produce the expected spirodiaza derivative 8, but it results in the formation of azaselenonium hydroxide **23** [54]. The azaselenonium cation contains one Se–N bond and the compound is stabilized by a noncovalent interaction between the selenium atom and the carbonyl oxygen atom of the other amide moiety (**Figure 6**).

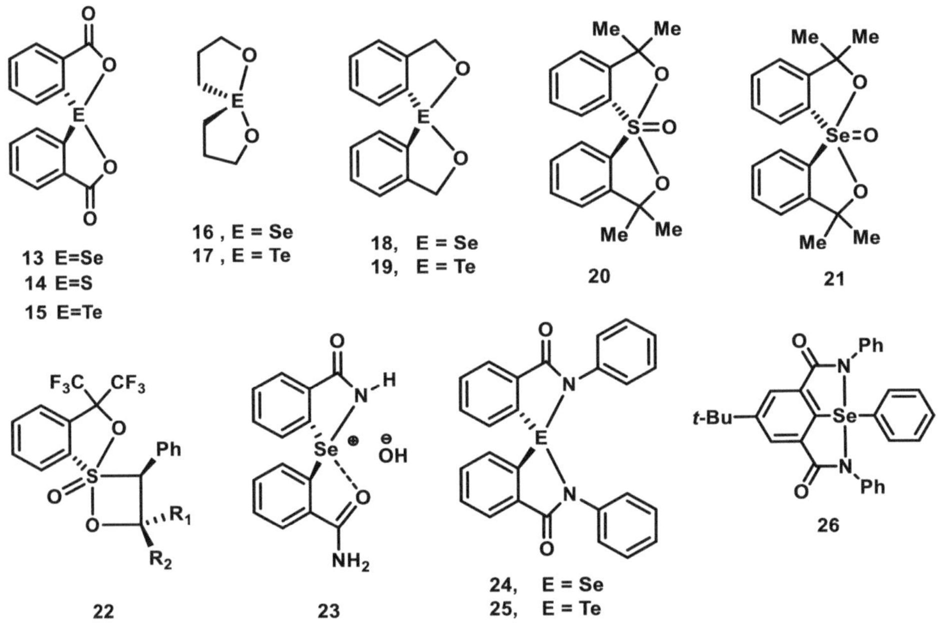

Figure 6.
Some representative examples of stable spirochalcogenuranes [47, 50, 53, 55].

In continuation selenium compounds in 2004 Back and co-workers reported for the first time that spirodioxyselenurane and its tellurium analogue exhibit very good antioxidant activity by mimicking the glutathione peroxidase (GPx) enzyme which protects the organism from oxidative damage by catalyzing the reduction of peroxides using thiol as the cofactor [47, 56]. Subsequently, a series of spirodioxyselenuranes with different stereochemistry and ring size has been reported [50, 54]. Very recently, we have reported the first example of a hydrolytically stable spirodiazaselenurane **24** and its tellurium analogue **25** bearing two nitrogen substituents in the apical positions [57]. In continuation of our research work on selenium compounds we have synthesized different substituted spirodiazaselenurane and its analogues and its Antioxidant activity [58, 59]. Very recently, Singh and co-workers have synthesized and characterized a new pincer type bicyclic diazaselenurane **26** where the two amide groups are present in the same phenyl ring forming the bicycles [55].

3.2 Glutathione peroxidase (GPx) activity

Back and co-workers reported that spirodioxyselenurane **11** exhibit excellent antioxidant property by mimicking glutathione peroxidise enzyme [47]. The effect of different substituents attached to the nitrogen atom was one of the objectives of this study to understand the antioxidant activity of selenides and spirodiazaselenuranes. Therefore, the GPx-like catalytic activity of compounds **27** was studied using glutathione (GSH) as thiol cofactor and hydrogen peroxide (H_2O_2) as substrate [57, 58]. The reduction of H_2O_2 by the selenides mechanism may involve a redox shuttle between the selenides and spirodiazaselenuranes via the corresponding intermatidates selenoxides (**Figure** 5). As previously described, the reactions of compounds **27** with H_2O_2 produce the corresponding selenoxides **28** which upon elimination of a water molecule generate the spirodiazaselenuranes **29**. Selenides **27** regenerate by GSH due to the reductive cleave of the Se-N bonds in compounds **30** (Path A, **Figure 7**). This pathway is particularly favored at higher concentrations peroxide.

The nucleophilic attack of the thiol at the selenium center may produce the intermediates **30** which upon reaction with GSH can regenerate the selenides at higher concentrations of GSH. The nucleophilic attack of GSH at the selenium center is expected to favor due to noncovalent interactions between the selenium and one of the carbonyl oxygen. It should be noted that the mechanism shown in **Figure 7** is different from that of GPx and other diselenide-based mimetics that utilize a selenol moiety for the reduction of peroxides.

3.3 Mechanism of spirocyclization

Dctailcd mcchanistic studics of spirodiazasclcnurancs and structural characterization were carried out by using ^{77}Se NMR spectroscopy. It was observed in the previous studies the presence of aromatic substituents cyclization process is very rapid at room temperature (25°C) on the nitrogen atoms as apical position [58]. To detect intermediates at room temperature (25°C) the cyclizations of the selenides to

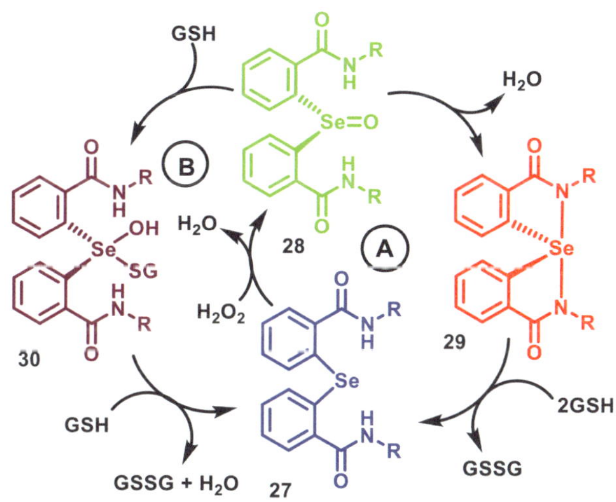

Figure 7.
Proposed mechanism for GPx activity of compounds **27–30** *[57, 58].*

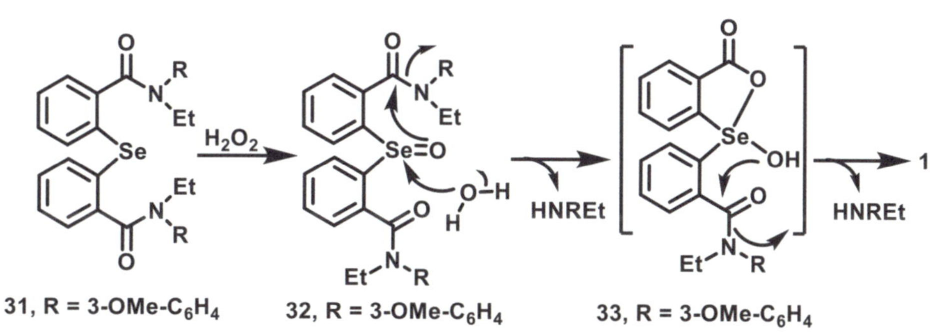

Figure 8.
Formation of spirodioxyselenurane from diaryl selenide by an oxidation-elimination mechanism [58].

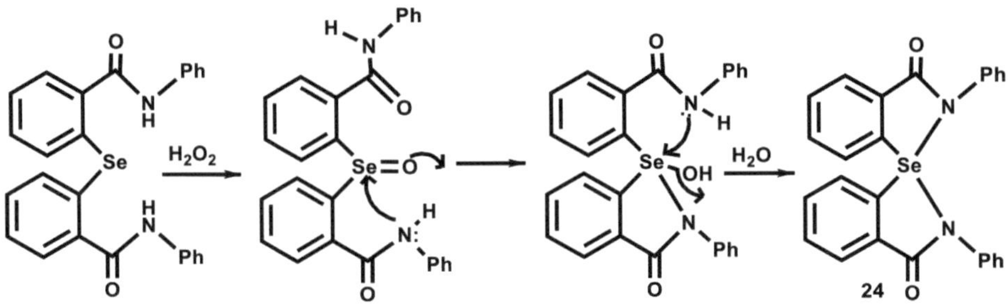

Figure 9.
Stepwise mechanism for the formation of Spirodiazachalcogenuranes from diaryl selenide [58, 60].

the corresponding spiro compounds were too fast. However, the formation of selenoxide could be observed when the cyclization is blocked by replacing the N-H moiety in compound **29** with an N-Et group. Therefore, the oxidation of selenide **29** by H_2O_2 produced the selenoxide **30** with very good yield.

However, it was observed when a solution of compound **31** in acetonitrile with H_2O_2 was kept for a week, formation of the corresponding spirodioxyselenurane **32**. The mechanism for the formation of **33** may proceed via the initial attack of a water molecule at the selenium center in compound **31** followed by a nucleophilic attack of the selenium-bound oxygen atom at the carbonyl carbon of one of the amide moieties, leading to the formation of an intermediate **32** (**Figure 8**) [59]. Although the formation of a spirodioxyselenurane has been proposed for a selenide having an *N*-methyl-*N*-phenylamide moiety [55, 58, 59] the conversion of **31** into **33** suggests that such mechanism can be generalized for diaryl selenides having different substituents on the amide nitrogen atom as mentioned in **Figure 9**.

4. Conclusions

In conclusion, a series of ebselen, diaryl selenides and spirodiazaselenuranes and its glutathione peroxidase (GPx) activity were discussed. A detailed mechanistic study suggests that the spirocyclization occurs *via* the formation of selenoxide intermediates. The glutathione peroxidase (GPx) mimetic activity of the selenides and the spirodiazaselenuranes indicates that the substituents attached to nitrogen atom have significant effect on the activity. Therefore, the selenoxide intermediates involved in the cyclization process could be isolated at the room temperature when it reacts with methyl halide. The comparison of the GPx-like activity showed that the antioxidant activity of diaryl selenids shows significant antioxidant activity due to oxidation of selenium and followed by the addition of H_2O_2 leads to spiro-compounds. Therefore, these compounds can provide a better protection against reactive oxygen species like H_2O_2 and peroxynitrite.

Acknowledgements

This study was supported by the Department of Science and Technology (DST), New Delhi. DSL thanks the UGC for the award of Dr. D. S. Kothari postdoctoral research fellowship and Department of Inorganic and Physical Chemistry, Indian Institute of Science, Bangalore. In addition author thanks to Prof. CNR Rao Research Centre Basaveshwar Science College, Bagalkot for support and encouragements.

Author details

Devappa S. Lamani
Department of Chemistry, Basaveshawar Science College, Bagalkot, Rani Channamma University, Belagavi, Karnataka, India

*Address all correspondence to: kattimani.devaraj67@gmail.com

References

[1] Berzelius JJ. Undersökning af enny Mineral-kropp, funnen ide orenare sorterna af detvid Fahlun tillverkade svafle, Discovery of Selenium. Afhandlingar i Fysik, Kemi och Mineralogi. 1818;**6**:42

[2] Löwig CJ. *Poggendorff's* Ueber schwefelwasserstoff-und selenwasserstoffäther (about hydrogen sulfide and selenium hydrogen ether). Annals of Physics. 1836;**37**:552

[3] Rathke B. Ueber das Selenmercaptan. Justus Liebigs Annalen der Chemie. 1869;**152**:211

[4] Hutton JG. The correlation of certain lesions in animals with certain soil types. Journal of the American Chemical Society. 1931;**23**:1076

[5] Franke KW. A toxicant occurring naturally in certain samples of plant foodstuffs. I. Results obtained in preliminary feeding trials. The Journal of Nutrition. 1934;**8**:597

[6] Franke KW, Potter VR. A new toxicant occurring naturally in certain samples of plant foodstuffs. IX. Toxic effects of orally ingested selenium. The Journal of Nutrition. 1934;**8**:615

[7] Franke KW, Painter EP. A new toxicant occurring naturally in certain samples of plant foodstuffs. XIV. The effect of selenium containing foodstuffs on growth and reproduction of rats at various ages. The Journal of Nutrition. 1936;**10**:599

[8] Pinsent J. The need for selenite and molybdate in the formation of formic dehydrogenase by members of the *coli-aerogenes* group of bacteria. The Biochemical Journal. 1954;**57**:10

[9] Andreesen R, Ljungdahl L. Formate dehydrogenase of *Clostridium thermoaceticum*: Incorporation of Selenium-75, and the effects of selenite, Molybdate, and tungstate on the enzyme. Journal of Bacteriology. 1973;**116**:867

[10] Turner DC, Stadtman TC. Purification of protein components of the clostridial glycine reductase system and characterization of protein a as a selenoprotein. Archives of Biochemistry and Biophysics. 1973;**154**:366

[11] Flohe L, Günzler EA, Schock HH. Glutathione peroxidase-selenoenzyme. FEBS Letters. 1973;**32**:132

[12] Boyington JC, Gladyshev VN, Khangulov SV, Stadtman TC, Sun PD. Crystal structure of formate dehydrogenase H: Catalysis involving Mo, molybdopterin, selenocysteine, and an Fe4S4 cluster. Science. 1997;**275**:1305

[13] Wilting R, Schorling S, Persson BC, Bock A. Selenoprotein synthesis in archaea: Identification of an mRNA element of *Methanococcus jannaschii* probably directing selenocysteine insertion. Journal of Molecular Biology. 1997;**266**:637

[14] Garcin E, Vernede X, Hatchikian EC, Volbeda A, Frey M, Fontecilla-Camps JC. The crystal structure of a reduced[NiFeSe] hydrogenase provides an image of the activity catalytic center. Structure. 1999;7:557

[15] Pfeiffer M, Bingemann R, Klein A. Fusion of two subunits does not impair the function of a [NiFeSe]-hydrogenase in the archaeon *Methanococcus voltae Eur*. Journal of Biochemistry. 1998;**256**:447

[16] Wagner M, Sonntag D, Grimm R, Pich A, Eckerskorn C, Söhling B, et al. Substrate-specific selenoprotein B of glycine reductase from *Eubacterium*

acidaminophilum. European Journal of Biochemistry. 1999;**260**:38

[17] Behne D, Kyriakopoulos A, Meinhold H, Köhrle J. Identification of type I iodothyronine 5′-deiodinase as a selenoenzyme. Biochemical and Biophysical Research Communications. 1990;**173**:1143

[18] Arthur JR, Nicol F, Beckett GJ. Hepatic iodothyronine 5′-deiodinase. The role of selenium. Biochemical Journal. 1990;**272**:537

[19] Davey JC, Becker KB, Schneider MJ, Germain DL, Galton VA. Cloning of a cDNA for the type II Iodothyronine Deiodinase. The Journal of Biological Chemistry. 1995;**270**:26786

[20] Croteau W, Whittemore SK, Schneider MJ, Germain DL. Cloning and expression of a Cdna for a mammalian type III Iodothyronine Deiodinase. The Journal of Biological Chemistry. 1995;**270**:16569

[21] Lescure A, Gautheret D, Carbon P, Krol A. Novel selenoproteins identified in silico and in viva by using a conserved RNA structural motif. The Journal of Biological Chemistry. 1999;**274**:38147

[22] Tamura T, Stadtman TC. A new selenoprotein from human lung adenocarcinoma cells: Purification, properties, and thioredoxin reductase activity. Proceedings of the National Academy of Sciences of the United States of America. 1996;**93**:1006

[23] Lee SR, Kim JR, Kwon KS, Yoon HW, Leveine RL, Ginsburg A, et al. Molecular cloning and Charactrisation of a Mitochandrial Selenocysteine-containing Thioredoxin Reductase from rat liver. The Journal of Biological Chemistry. 1999;**274**:4722

[24] Mustacich D, Powis G. Thioredoxin reductase. The Biochemical Journal. 2000;**346**:1

[25] Watabe S, Makino Y, Ogawa K, Hiroi T, Yamamoto Y, Takahashi SY. Mitochondrial thioredoxin reductase in bovine adrenal cortex its purification, properties, nucleotide/amino acid sequences, and identification of selenocysteine. European Journal of Biochemistry. 1999;**264**:74

[26] Mills GC. Hemoglobin catabolism I. Glutathione peroxidase, an erythrocyte enzyme which protects hemoglobin from oxidative breakdown. The Journal of Biological Chemistry. 1957;**229**:189

[27] Motsenbocker MA, Tappel AL. Effect of dietary selenium on plasma selenoprotein P, selenoprotein P1 and glutathione peroxidase in the rat. The Journal of Nutrition. 1984;**114**:279

[28] Ursini F, Maiorino M, Valente M, Ferri L, Gregolin C. Purification from pig liver of a protein which protects liposomes and biomembranes from peroxidative degradation and exhibits glutathione peroxidase activity on phosphatidylcholine hydroperoxides. Biochimica et Biophysica Acta. 1982;**710**:197

[29] Takahashi K, Avissar N, Whittin J, Cohen H. Purification and characterization of human plasma glutathione peroxidase: A selenoglycoprotein distinct from the known cellular enzyme. Archives of Biochemistry and Biophysics. 1987;**256**:677

[30] Chu FF, Doroshow JH, Esworthy RS. Expression, characterization, and tissue distribution of a new cellular selenium-dependent glutathione peroxidase, GSHPx-GI. Journal of Biological Chemistry. 1993;**268**:2571

[31] Gromer S, Eubel JK, Lee BL, Jacob J. Human selenoproteins at a glance.

Cellular and Molecular Life Sciences. 2005;**62**:2414

[32] Flohé L. Glutathione peroxidase brought into focus in into focus. In: Pryor WA, editor. Free Radicals in Biology. Vol. 5. New York: Academic Press; 1982. p. 223

[33] Maiorino M, Roveri A, Coassin M, Ursini F. Kinetic mechanism and substrate specificity of glutathione peroxidase activity of ebselen (PZ51). Biochemical Pharmacology. 1988;**37**:2267

[34] Epp O, Ladenstein R, Wendel A. The refined structure of the selenoenzyme Gultathione peroxidase at 0.2-nm resolution. European Journal of Biochemistry. 1983;**133**:51-69

[35] Nogueira CW, Rocha JBT. Organoselenium and organotellurium compounds: Toxicology and pharmacology. In: Rappoport Z, editor. The Chemistry of Organic Selenium and Tellurium Compounds. Vol. 1, 2, Part-II, Chapter 21. New York: Wiley; 2012

[36] Mukherjee AJ, Zade SS, Singh HB, Sunoj RB. Organoselenium chemistry: Role of intramolecular interactions. Chemical Reviews. 2010;**110**:4357

[37] Müller A, Cadenas E, Graf P, Sies H. A novel biologically active seleno-organic compound-1: Gultathione peroxidase-like activity in vitro and antioxidant capacity of PZ-51(Ebselen). Biochemical Pharmacology. 1984;**33**:3235

[38] Sies H, Matsumoto H. Ebselen as a glutathione peroxidase mimic and as a scavenger of peroxynitrite. Advances in Pharmacology. 1997;**38**:229-246

[39] Fong MC, Schiesser CH. Reactions of 2,2′-diselenobis(N-alkylbenzamides) with peroxides: A free-radical synthesis of *Ebselen* and related analogues. Tetrahedron Letters. 1995;**36**:7329

[40] Mugesh G, Singh HB. Synthetic organoselenium compounds as antioxidants: Glutathione peroxidase activity. Chemical Society Reviews. 2000;**29**:347

[41] Mugesh G, Panda A, Singh HB, Punekar NS, R. J. Butcher gultathione peroxidase-like activity of diaryl diselenides-a mechanistic study. Journal of the American Chemical Society. 2001;**123**:839

[42] Bhabak KP, Mugesh G. Synthesis, characterization and antioxidant activity of some ebselen analogues. Chemistry—A European Journal. 2007;**13**:4594

[43] Bhabak KP, Mugesh G. Antioxidant activity of the anti-inflammantry compound ebselen; a reversible cyclisation pathway via selenenic and seleninic acid intermadates. Chemistry—A European Journal. 2008;**14**:8640

[44] Bhabak KP, Mugesh G. Amide-based glutathione peroxidase mimics: Effect of secondry and tertiary amid sustituenets on antioxidant activity. Chemistry, an Asian Journal. 2009;**4**:974

[45] Wirth T, Fragale G, Spichty M. Mechanistic course of the asymmetric Methoxyselenenylation reaction. Journal of the American Chemical Society. 1998; **120**:3376

[46] Back TG, Moussa Z. Remarkable activity of a novel cyclic seleninate ester as glutathione peroxidase mimetic and its facile in situ generation from allyl-3-hydroxypropyl selenide. Journal of the American Chemical Society. 2003;**125**:13455

[47] Back TG, Moussa Z, Parvez M. The exceptional glutathione peroxidase-like activity of Di(3-hydroxypropyl) Selenide and the unexpected role of a novel Spirodioxaselenanonane intermediate in

the catalytic cycle. Angewandte Chemie, International Edition. 2004;**43**:1268

[48] Lesser R, Weiss R. Über Selenoxanthon und Selenoxanthon-carbonsäure. Über selenhaltige aromatische Verbindungen Über *Ber.* Dtsch. Chemische Berichte. 1914;**47**:2510-2525

[49] Kapovits I, Kalman A. Formation and structure of a four-co-ordinate Organo-Sulphur (IV) compound. Chemical Communications. 1971;**12**:649-650

[50] Takaguchi Y, Furukawa N. First synthesis and structural determination of 1, 1′-spirobis(3H-2, 1-benzoxatellurole)-3, 3′-dione ([10-Te-4(C2O2)]). Heteroatom Chemistry. 1995;**6**:481-484

[51] Bhabak KP, Mugesh G. Functional mimics of glutathione peroxidase: Bioinspired synthetic antioxidants. Accounts of Chemical Research. 2010;**43**: 1408-1419

[52] Zhang Z, Takahashi S, Saito S, Koizumi T. First synthesis and stereochemistry of enantiomerically pure spiroselenurane and spirotellurane using the 2-exo-hydroxy-10-bornyl group as a chiral ligand. Tetrahedron: Asymmetry. 1998;**9**:3303-3317

[53] Zhang J, Saito S, Koizumi T. Acidic and basic hydrolysis of an optically pure Spiro-λ4-sulfurane: Completely opposite Stereochemical outcome. Journal of the American Chemical Society. 1998;**120**:1631-1632

[54] Kapovits I, Rabai J, Szabó D, Czako K, Kucsman A, Argay G, et al. Diaryldiacyloxyspirosulfuranes. Part 3. Sulfuranes with five-, six- and Sevenmembered Spirorings: Syntheses and molecular structures. Journal of Chemical Society, Perkin Transactions. 1993;**2**:847-853

[55] Selvakumar K, Singh HB, Goel N, Singh UP, Butcher RJ. Synthesis and structural characterization of pincer type bicyclic diacyloxy- and diazaselenuranes. Dalton Transactions. 2011;**40**:9858-9867

[56] Jacob C, Giles GI, Giles NM, Sies H. Sulfur and selenium: The role of oxidation state in protein structure and function. Angewandte Chemie, International Edition. 2003;**42**:4742-4758

[57] Back TG, Kuzma D, Parvez M. Aromatic derivatives and tellurium analogues of cyclic Seleninate esters and spirodioxyselenuranes that act as glutathione peroxidase Mimetics. The Journal of Organic Chemistry. 2005;**70**:9230-9236

[58] Sarma BK, Manna D, Minoura M, Mugesh G. Synthesis, structure, Spirocyclization mechanism, and glutathione peroxidase-like antioxidant activity of stable spirodiazaselenurane and spirodiazatellurane. Journal of the American Chemical Society. 2010;**132**:5364-5374

[59] Lamani DS, Bhowmick D, Mugesh G. Substituent effects on the stability and antioxidant activity of spirodiazaselenuranes. Molecules. 2015;**20**:12959-12978

[60] Lamani DS, Bhowmick D, Mugesh G. Spirodiazaselenuranes: Synthesis, structure and antioxidant activity. Organic & Biomolecular Chemistry. 2012;**10**:7933-7943

Section 2

Solar Cells and Batteries

Chapter 5

Two-Dimensional Transition Metal Dichalcogenide as Electron Transport Layer of Perovskite Solar Cells

Akrajas Ali Umar, Nurul Ain Abd Malek, Nabilah Alias and Abang Anuar Ehsan

Abstract

Conventional perovskite solar cells utilize a combination of a compact and mesoporous layer of TiO_2 or SnO_2 as the electron transport layer. This structure is vulnerable to massive loss of photogenerated carriers due to grain boundary resistance in the layer. In this chapter, we will discuss a potential electron transport layer that might drive higher power conversion efficiency, i.e., thin and single-crystalline 2D transition metal dichalcogenide. Because of their ultimate thin structure, they facilitate rapid electron transport and enhanced carrier extraction in the solar cells device. We will also discuss the current state of the art of 2D transition metal dichalcogenide atomic layer application as an electron transport layer in the perovskite solar cells as well as our recent attempt in this field.

Keywords: 2D atomic layer, transition metal dichalcogenide, electron transport layer, perovskite solar cells

1. Introduction

Perovskite solar cells (PSCs) have received a great deal of attention in the past few decades due to their impressively high power conversion efficiency (PCE) [1]. To date, PCE as high as 25.6% has been successfully recorded. This performance has already been compared with the single-crystalline silicon solar cells system. With the advancement in the perovskite properties control, including the crystallinity properties, grain size, and stability properties, further improvement in the PCE is expected to be achieved soon. The continuous growth in the preparation of the high-performance charge selective layer in the perovskite solar cells further contributes to the rapid progress in the PCE improvement of the PSC [2].

Along with the transparent conducting electrode (TCE) and the top metal contact, a PSC device is composed of an electron-transport layer (ETL), an organometal-halide perovskite active layer, and a hole-transport layer (HTL). In these solar cells, the perovskite and its photoelectrical properties are the keys to the overall photovoltaic process. Its unique high-optical absorption constant drives massive photon absorption and exciton generation in the device. Despite this key fact, the carrier transport and interfacial charge transfer dynamics play another crucial factor for the generation of the overall PSC performance. These two parameters depend on the nature of the surface and the crystallinity properties of the charge-selective layers [3].

One of the serious problems in perovskite solar cell devices is the loss of charge carriers during the transport process in the carrier layer. This is because, the carrier layer has low crystallization, high grain boundary resistance as well as experiences loss of carrier charge during extraction to the outer electrode. The main factor of carrier charge lost during extraction to the outer electrode is due to the high interface resistance between the electrode and the carrier layer. Therefore, it is expected that when a carrier layer that has high crystallinity, very low thickness, and good coupling conditions with external electrodes is used, then the performance of the device will increase.

The electron transport layer (ETL), for example, TiO_2, and other semiconducting oxides, such as SnO_2, ZnO, have been widely applied in the perovskite solar cells fabrication. Despite the excellent performance demonstrated by them, this ETL suffers from large-density surface defects related to oxygen vacancy, particularly in the TiO_2 system. The defect from such vacancy causes immense trap-limited (Shockley-Read-Hull) transport in the extraction of the photogenerated carrier to the external electrode. This in many cases degrades the photovoltaic performance of the PSC up to a certain degree, reducing the power conversion efficiency of the device. Even though there exist several methods in the passivation of such defects, such as acid passivation, etc., the improvement is minute. In addition, this method may add additional resistance

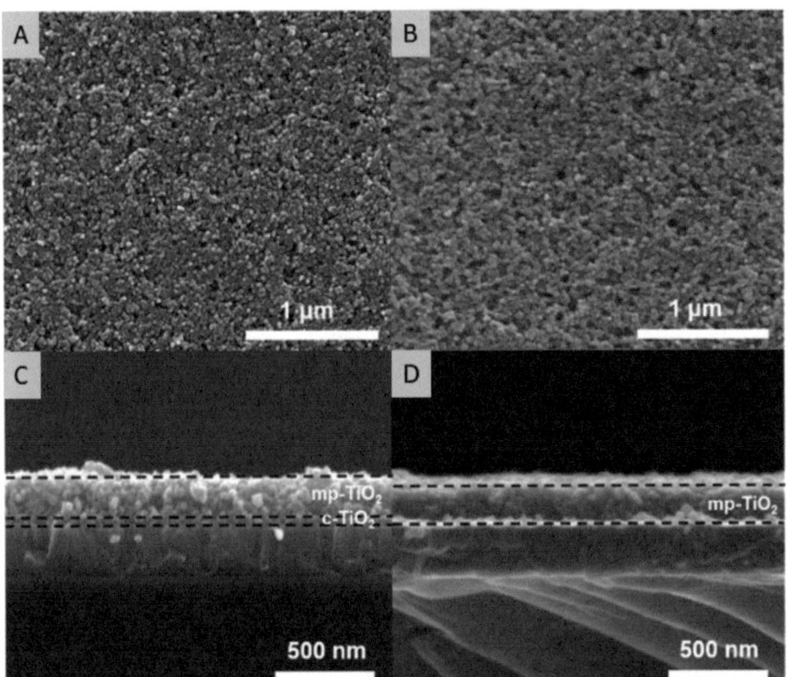

Figure 1.
Mesoporous TiO_2 ETL. (A and C) Top and side view of mesoporous TiO_2 layer on compact layer TiO_2. (B and D) Top and side of mesoporous TiO_2 layer. (Reprinted from [4]. © 2017 American Chemical Society).

to the photocarrier transport reducing the power conversion efficiency. Along with these crucial factors, the crystallinity properties of the ETL add an additional issue to the photocarrier transport dynamic in the device. As normal in the high-performance PSC fabrication, mesoporous TiO_2 or SnO_2 was used as ETL along with a compact layer of TiO_2 or SnO_2 (See **Figure 1**), [4]. As the figure reveals, the mesoporous layer is composed of a large number of interconnected small grain particles that produce grain boundary resistance due to lattice mismatch among the connected particles. This resistance should be massive due to their large-scale existence on the layer. This certainly complicates the transport of photogenerated electrons to the electrode layer, such as high internal resistance or radiationless recombination [5, 6]. Therefore, the selection of the right material for the carrier layer is important in determining the performance of a device. Such resistance boundary further augments the presence of mesoporous-compact layer interface resistance in the ETL system of the PSC. From this picture, we can estimate the loss would be suffered by the device during the photovoltaic process. This means that if such ETL is replaced with the single-crystalline ETL system, the performance of the perovskite solar cells can be improved.

2. Two-dimensional transition metals ETL

Recently, materials of two-dimensional (2D) dichalcogenide transition metals (TMDs), such as MoS_2, WS_2, TiS_2, CdTe, and others, have been used as carrier layers in perovskite solar cells due to their high charge carrier mobility, unique optoelectrical properties, large exciton binding energy, very fast interface charge transfer properties as well as excellent physical and chemical stability properties [7]. Their optoelectronic properties were also found to correspond to the number of layers, dopants, and strains (straining). The phenomenon of the massive charge transfer process in these van der Waals crystals driven by the collective motion of excitonic surfaces enables a high interfacial charge extraction and reduces charge recombination for an effective photovoltaic process [8]. One of the uniqueness of the TMDs layer is that it has an atomic-scale thickness (very thin) and has high crystallinity. With its planar (2D) structure, it will produce a strong coupling when grown on the electrode surface. Therefore, it has great potential for a carrier layer in perovskite solar cells.

Transition metal dichalcogenide (TMD) has the chemical formula of MX_2 where M is the transition metal from groups 4 to 10 in the periodic table system, and X is a chalcogen atom such as sulfur (S), selenium (Se), or tellurium (Te). **Figure 2** shows the typical structure of TMD. The structure has two layers of chalcogen that clamp a transition metal layer making this material have its uniqueness in electronic, opto-electronic properties, and chemical stability [10]. The electronic and optical properties of TMDs materials change significantly depending on the number of layers. For example, the MoS_2 band gap increases from 1.29 eV (multilayered MoS_2) to 1.59 eV (monolayer MoS_2), and also this bandgap changes from an indirect bandgap to a direct bandgap as the number of layers decreases [11].

As is well known, most of these 2D TMD materials have ambipolar properties that enable the materials to transport both electrons and holes [12]. In other words, this allows 2D TMDs material to be used as ETL or HTL in n-i-p or p-i-n perovskite solar cells. However, most perovskite solar cell applications use these 2D TMD materials as HTL. Only MoS_2 and TiS_2 have been used as ETLs and have successfully produced efficiencies as high as 13.14% and 18.79% [7, 13]. **Table 1** shows several PSC device structures utilizing TMD as ETL. Recently, there was a first simulation study on the

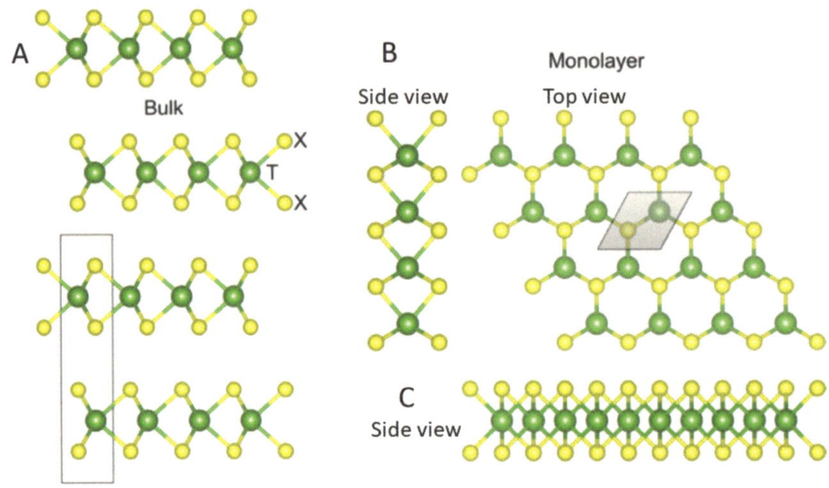

Figure 2.
Typical structure of transition metal dichalcogenide materials. (A) Typical layer stacking structure in bulk transition metal dichalcogenide structure. T and X represent the transition metal and chalcogen elements, respectively. (B) Top and side view of single-layer of TMD with 2H-phase. (C) Side view of single-layer TMD with 1T-phase. (Reprinted from [9]. © 2020, The Author(s)).

Material	Device structure	J_{sc} (mA cm^{-2})	V_{oc} (V)	*FF*	PCE (%)	Ref.
TiS_2	FTO/TiS_2/$MAPbI_3$/spiro-OMeTAD/Au	23.38	1.05	0.71	17.37	[14]
TiS_2	ITO/TiS_2/ $FA_xMA_{1-x}Br_xCl_yI_{1-x-y}$/spiro-OMeTAD/Ag	24.68	1.00	0.75	18.79	[7]
MoS_2	FTO/MoS_2/$MAPbI_3$/spiro-OMeTAD/Au	21.70	0.89	0.63	13.14	[15]
MoS_2	ITO/MoS_2/$Cs_x(MA_yFA_{1-y})_{1-x}Pb(I_zBr_{1-z})_3$/spiro-OMeTAD/Au	16.24	0.56	0.37	3.36	[16]
MoS_2/TiO_2	ITO/TiO_2/MoS_2/$MAPbI_3$/spiro-OMeTAD/Au	13.36	0.65	0.51	4.43	[17]
MoS_2/SnO_2	ITO/SnO_2/MoS_2/$FA_xMA_{1-x}Br_xCl_yI_{1-x-y}$/spiro-OMeTAD/Ag	24.57	1.11	0.79	21.73	[18]
MoS_2	Graphene/MoS_2/$MAPbI_3$/PTAA/Au	20.92	0.91	0.76	14.42	[19]
MoS_2	ITO/MoS_2/$MAPbI_3$/PCBM/Al	12.50	0.85	0.57	6.01	[20]
SnS_2	ITO/SnS_2/$MAPbI_3$/Spiro-OMeTAD/Au	23.70	0.95	0.61	13.63	[21]
SnS_2	ITO/SnS_2/$MAPbI_3$/Spiro-OMeTAD/Au	21.70	1.011	0.60	13.20	[22]

Table 1.
Photovoltaic parameters of perovskite solar cell devices using dichalcogenide transition metals (TMDs) as ETLs.

photoelectric properties of WS_2 as an ETL in perovskite solar cells reported with efficiencies as high as 25.70% [23]. By having high electron mobility as well as energy levels appropriate to the perovskite layer, the WS_2 atomic layer is expected to function as an ETL capable of producing high-performance perovskite solar cell devices.

2.1 TiS_2 electron transport layer

TiS_2 is one of the TMDC family that has been intensively studied recently due to its semi-metallic properties with low-bandgap value, i.e., 0.2 eV. With high electrical conductivity, i.e., 1×10^4 S m^{-1}, this material is potential as an electrode in many applications including lithium-ion batteries and solar cells. Despite its excellent electrical properties, the use of TiS_2 as independent electrode material in the application is limitedly demonstrated. It is mainly stacked with other materials such as MoS_2 [24] or TiO_2 to improve the properties in applications. For the case of MoS_2 stacked with TiS_2, the TiS_2 can form Schottky contact with MoS_2 with barrier height [24] between these two atomic layers can be varied by the doping type and concentration either in the MoS_2 or TiS_2 side (**Figure 3**). This certainly provides a wider opportunity to modify the electrical properties of the system for desired

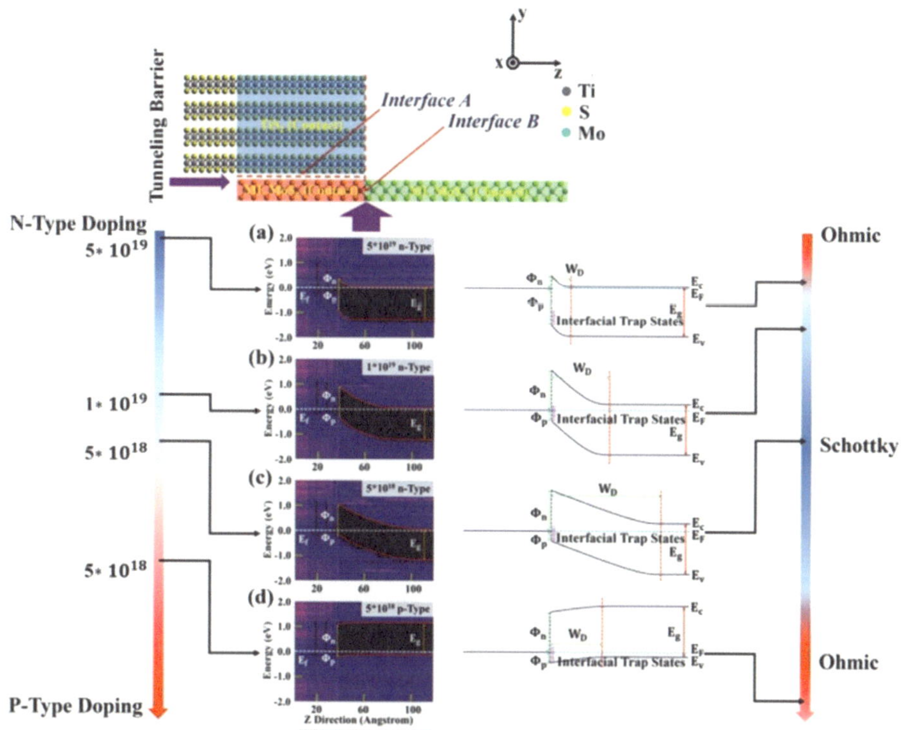

Figure 3.
PLDOS of TiS_2–MoS_2 (ML) FET-like junctions doped with different doping concentrations and the variation of band structure at interface B. a–d The doping concentrations are: $N = 5 \times 10^{19}$ cm^{-3}, $N = 1 \times 10^{19}$ cm^{-3}, $N = 5 \times 10^{18}$ cm^{-3}, and $P = 5 \times 10^{18}$ cm^{-3}. The thickness of TiS_2 is four layers. On the right side, the plot shows the variation of band structure under different doping concentrations. The scale bar is from 0.0 to 90.0 (1/eV). Interface A is the interface between TiS_2–MoS_2. (Reprinted from [24]. © 2020, The Author(s)).

performance in application. In the typical process, n-type-doped TiS_2–MoS_2 (ML) contacts exhibit a barrier height relatively larger, i.e., 1.0 eV below doping level degeneracy. Nevertheless, these n-type-doped contacts still have the potential as the switch in high-power as well as tunnel Schottky barrier MOSFETs. In contrary to the n-type doped system, the p-type-doped TiS_2–MoS_2 (ML) exhibits a zero barrier height at a particular doping concentration, i.e., 5×10^{18} cm^{-3}. Under this condition, the depletion region width is zero and the band becomes flat, revealing that the contact is ohmic and the barrier height is small. These results reveal the unique unusual interfacial properties arising from this ultimate thin contact that promise a special function in the application. This phenomenon could be the driving factor for an efficient photocarrier extraction in the perovskite solar cells using ETL modified with MoS_2 or TiS_2 atomic layer.

For example, in the perovskite solar cells system with SnO_2 ETL (**Figure 4**), there is an increase in the energy band alignment between the ETL and perovskite layer when the 2D TiS_2 is attached to the surface of SnO_2 [18]. The conduction band level

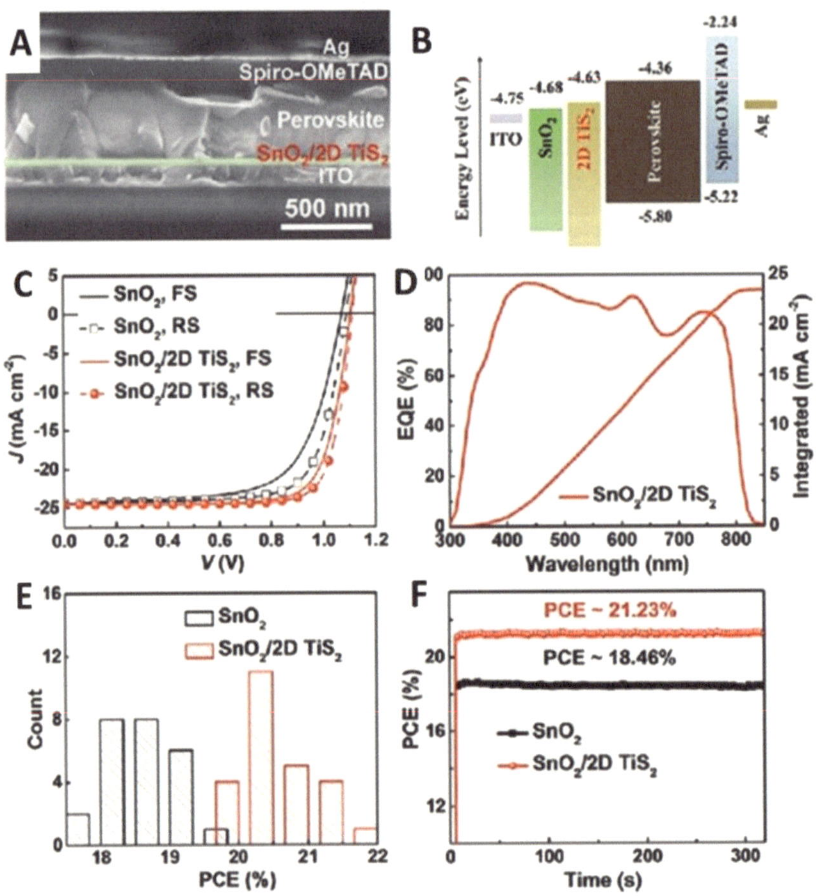

Figure 4.
(A) Cross-sectional SEM image of the PSC. (B) The energy level diagram. (C) Representative J-V curves of the PSCs with SnO_2 or SnO_2 /2D TiS_2 as ETLs. (D) EQE curve and integrated current density of the PSC with SnO_2 /2D TiS_2 as the ETL. (E) Histogram of the PCE of PSCs with SnO_2 and SnO_2 /2D TiS_2 as ETLs analyzed from 25 cells. (F) Steady-state efficiency of the PSCs with SnO_2 and SnO_2/2D TiS_2 as ETLs measured under constant voltages of 0.86 V and 0.92 V, respectively. (Reprinted from [18]. © 2019 Royal Society of Chemistry).

of ETL (SnO_2) reduced from 4.68 to 4.63 eV in the presence of 2D TiS_2. This has narrowed the offset energy between the ETL and perovskite (conduction band level at 4.36 eV). As the result, the photogenerated carrier extraction becomes enhanced, improving the photocurrent and the power conversion efficiency. As shown in **Figure 4C–4F**, the power conversion efficiency increases from 19.65% to 21.73% when the SnO_2 ETL is modified with the 2D TiS_2 atomic layer. The nature of interfacial photocarrier dynamic improvement in the presence of the 2D TiS_2 atomic layer can be seen from the increase of the V_{oc}, FF, and the IPCE of the device. This process is also reflected by the decrease in the device hysteresis and the improvement of the stability properties.

Figure 5 explains in detail how the photocarrier dynamic in the device was impressively modified in the presence of a 2D TiS_2 atomic layer on the surface of SnO_2 ETL. As presented, the photocurrent is enhanced impressively. This is the result of

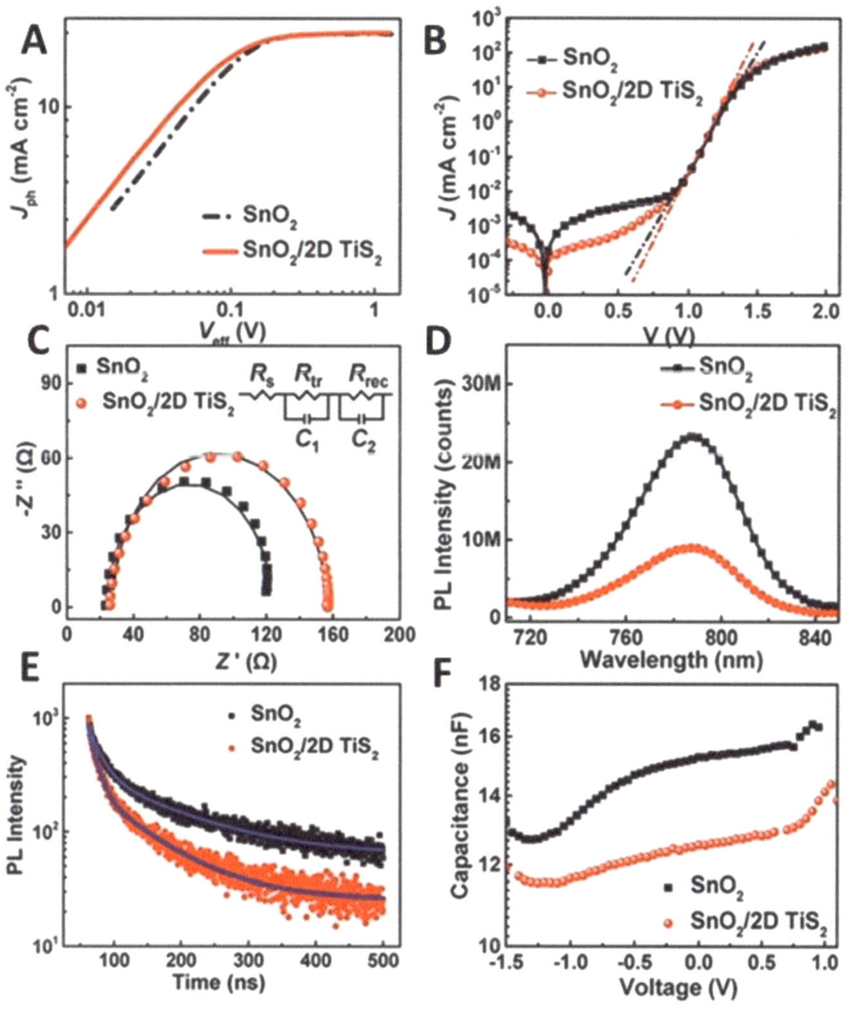

Figure 5.
Comparison of SnO_2 and SnO_2/2D TiS_2 as ETLs in PSCs: (A) J_{ph}-V eff curves; (B) J-V curves in the dark (the dash-dot lines represent the fitting lines); (C) Nyquist plots; (D) steady-state PL spectra; (E) transient PL spectra; (F) C-V characteristics. (Reprinted from [18]. © 2019 Royal Society of Chemistry).

enhanced interfacial charge transfer as indicated by the transient and steady-state photoluminescence analysis result, which is also supported by the electrochemical impedance spectroscopy result, showing decrease in the interfacial charge transfer resistance in the device.

We also in our recent result have coupled the TiS_2 atomic layer on top of the TiO_2 surface to compensate for surface defect due to the oxygen vacancy, enhancing the interfacial charge transfer and transport dynamic when applied as ETL in perovskite solar cells [25]. The perovskite solar cells' performance improves from 18.02 to 18.73% (**Figure 6**). Electrochemical impedance analysis revealed that there is an improvement as high as 13% in interfacial charge transfer in the ETL with 2D TiS_2 and 43% improvement in the charge recombination resistance (**Figure 7A**). The latter is verified by the increase in the photocurrent (**Figure 7B**) and the decrease in the leakage

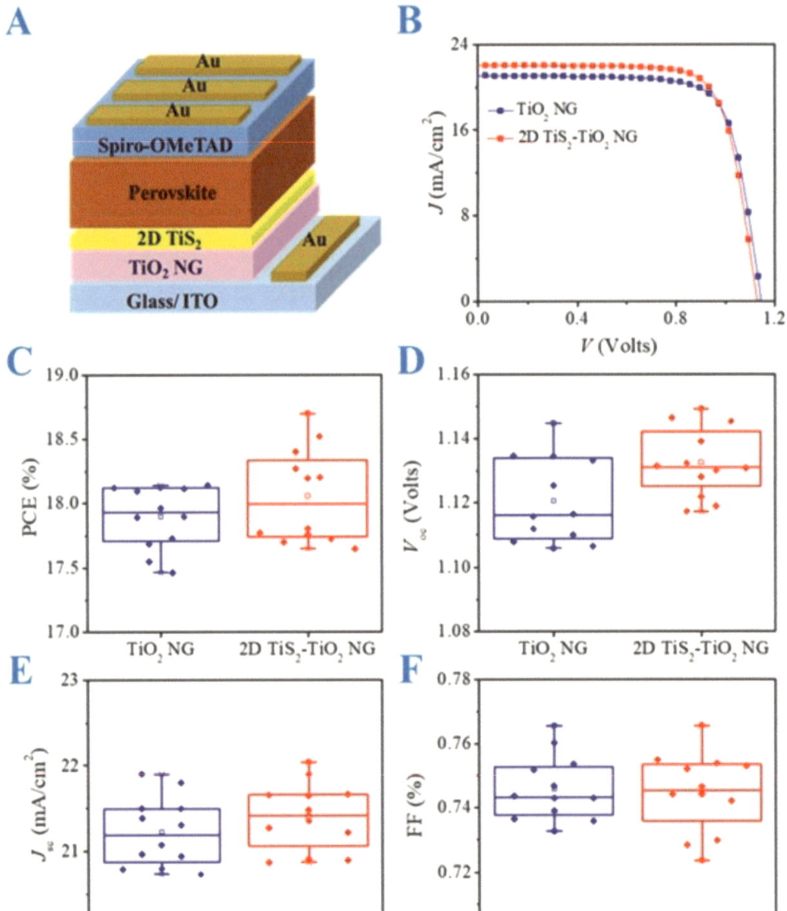

Figure 6.
Photovoltaic performance of the 2D TiS_2-TiO_2 NG and TiO_2 NG-based PSC. (A) Schematic structure of 2D TiS_2-TiO_2 NG-based PSC. (B) J-V curves of the champion device. (C-F) The comparison of the photovoltaic parameters, i.e., PCE, V_{oc}, J_{sc}, and FF, for the two devices. (Reprinted from [25]. © 2021 The American Chemical Society).

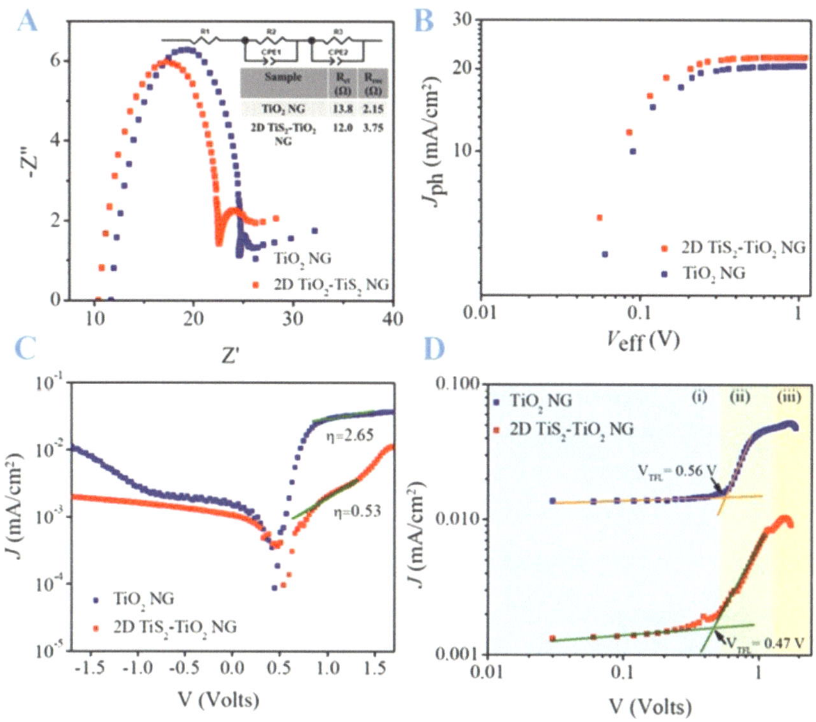

Figure 7.
Photoelectrical properties of the PSC device. (A) Electrochemical impedance spectra and equivalent circuit of the device. (B) Photogenerated current of the PSC device (J_{ph}-V_{eff} curve). (C) Semilog J-V curve of the PSC in the dark. The green lines represent the fitting line. (D) Double log J-V curve in the dark for photoelectrical dynamic in the device. Three distinct regimes of (i) the ohmic response, (ii) filled trap transition, and (iii) SCLC are shown by different colored regions. (Reprinted from [25]. © 2021 The American Chemical Society).

current of the device when 2D TiS_2 passivates the TiO_2 surface (**Figure 7C**). We can relate this process to the reduction in the trap density in the device as shown by the value of V_{TFL} of the double-log *J-V* curve as depicted in **Figure 7D** where the V_{TFL} value depends linearly with the trap density in the device.

2.2 MoS_2 electron transport layer

MoS_2 atomic layer is the most studied TMD system because of its excellent optical and electrical properties [26–28] and has been used widely in perovskite solar cells as a hole-transport layer (HTL) and an electron-transport layer (ETL) [11, 15, 26] in the form of colloidal or flakes thin film [15, 28–30]. **Table 1** lists down several perovskite solar cells using MoS_2 as ETL with a particular device configuration. For example, Singh, Giri, et al. [13] have obtained power conversion efficiency as high as 13.2% from PSC devices using MoS_2 material as ETL. In this study, they synthesized the MoS_2 film directly on FTO substrate using microwave irradiation-assisted reduction method. It is found that the efficiency obtained by MoS_2 material is close to the efficiency value obtained from TiO_2 and SnO_2 material making MoS_2 material comparable to other ETL materials. Abd Malek et al. [16] have also developed different structures of MoS_2 ETL on the ITO substrate. Instead of colloidal or flake structured film, an ultrathin layer of MoS_2 prepared from ultrasonic spray pyrolysis was fabricated to obtain its functionalities as ultrathin ETL in the PSC device. The result showed that the PCE device performance depended on the condition during the preparation of the MoS_2 atomic layer, particularly the substrate temperature. It is demonstrated that substrate temperature of 200°C is suitable for growing high-quality MoS_2 atomic layer on ITO surface, thus, optimizing the power conversion efficiency of the PSC (**Figure 8**). This MoS_2 thin-film-based device as ETL has shown high-stability properties where its efficiency can be maintained as much as 90.24% of the original efficiency after 80 s exposure continuously under simulated solar light illumination (AM1.5).

In addition to being used singly in the ETL, TMD materials can also be combined with other organic or inorganic electron transport materials to form electron transport materials. For example, Ahmed et al. [31] have added a MoS_2 layer on top of the TiO_2 layer to be used as ETL in perovskite solar cells. The use of MoS_2/TiO_2 as ETL has successfully increased the efficiency of the device by 16% higher than the device

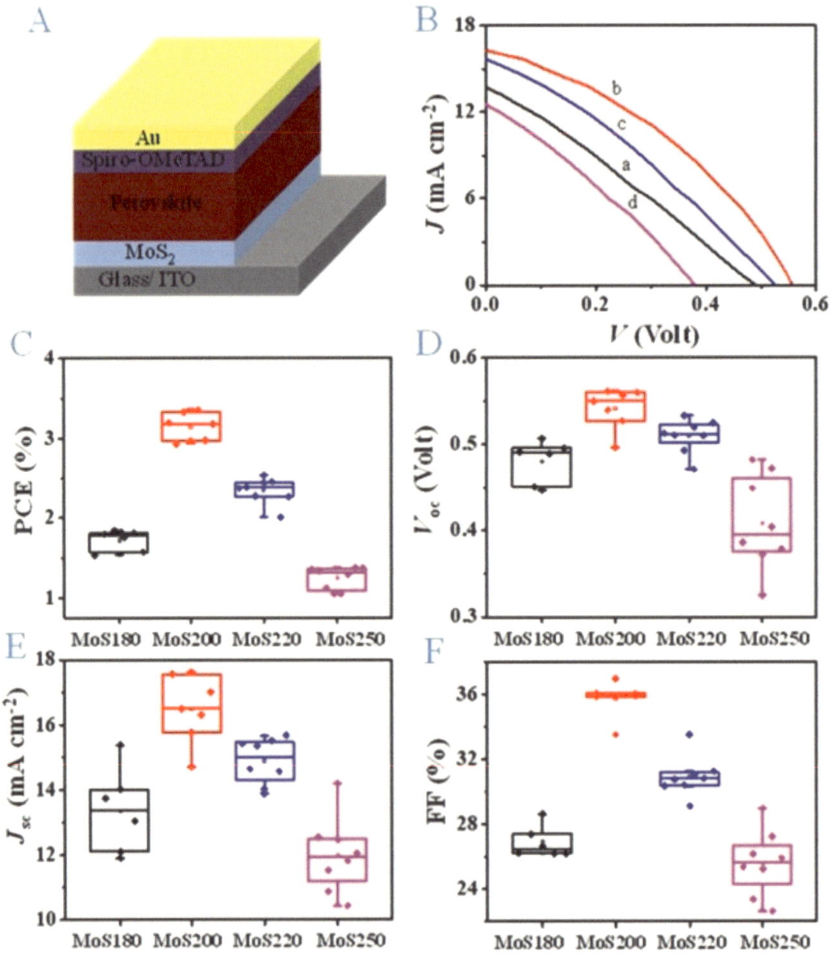

Figure 8.
The photovoltaic parameter for MoS_2 as ETL in PSC. (A) Schematic structure of the PSC device. (B) The J-V curves for the champion device, (C-F) PCE, V_{oc}, J_{sc}, and FF of the MoS_2 based PSC devices with different substrate temperatures, namely MoS180 (a), MoS200 (b), MoS220 (c), and MoS250 (d). (Reprinted from [16]. © 2020 Elsevier).

that only uses TiO_2 as ETL. Similarly, Huang et al. [18] have successfully produced an n-i-p type plane device using SnO_2 and 2D TiS_2 as ETL. High efficiency was recorded by this group, which was as high as 21.73% with a relatively small hysteresis value. The increase in efficiency in this device is due to the matching of the ETL energy level and the appropriate perovskite layer as well as the lack of electron trap density in the ETL.

2.3 WS_2 electron transport layer

Tungsten disulfide (WS_2) share common basic properties of TMD with other systems, such as high-mobility properties, unique optoelectronic properties, large exciton-binding energy, and good physical and chemical stability as well as ambipolar properties [11]. In addition, WS_2 has an energy level that is suitable for the perovskite layer of three types of cations (**Figure 9**) and can be easily synthesized by the ultrasonic spray pyrolysis method. WS_2 also has high stability as well as having fast interface charge transfer properties [32]. Among the available 2D TMD, the energy band structure of WS_2 is a much better match with the common perovskite of $MAPbI_3$ (**Figure 10**). Furthermore, it also has a relatively larger bandgap if compared with the other system in this class of materials, promising facile excitonic separation during the photovoltaic process and producing better power conversion efficiency.

Recently, we have realized the PSC device utilizing the WS_2 layer as ETL and evaluated how the number of layers of WS_2 influences the carrier dynamic in the device [5]. We prepared the WS_2 atomic layer via ultrasonic spray pyrolysis. **Figure 11** shows a schematic diagram of the 2D atomic layer preparation. A modified commercially available ultrasonic spray system (Daiso, Japan) was used. A homemade solution container was placed on the top of the ultrasonic membrane of the system (**Figure 11**). Ultrasmall solution precursor mist can be produced from the process and fall on the ITO substrate surface that is positioned approximately 5 cm below the membrane. The temperature of the substrate was set at 350°C.

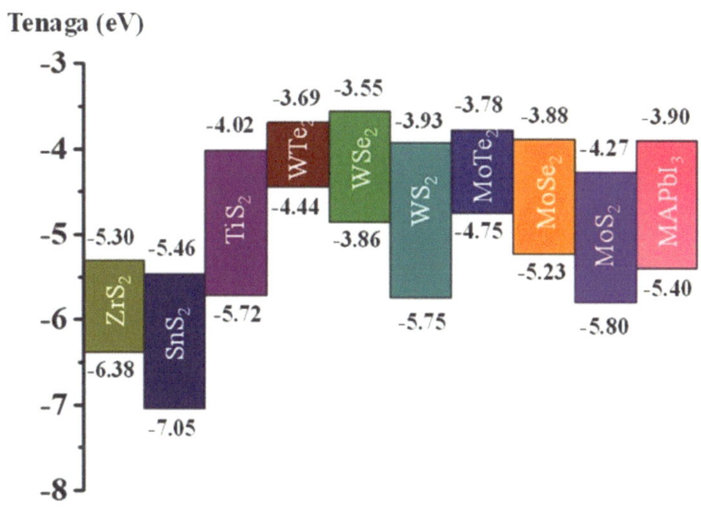

Figure 9.
Energy levels of dichalcogenide transition metal materials (TMDs) as ETLs and MAPbI3 as perovskite layers in perovskite solar cells.

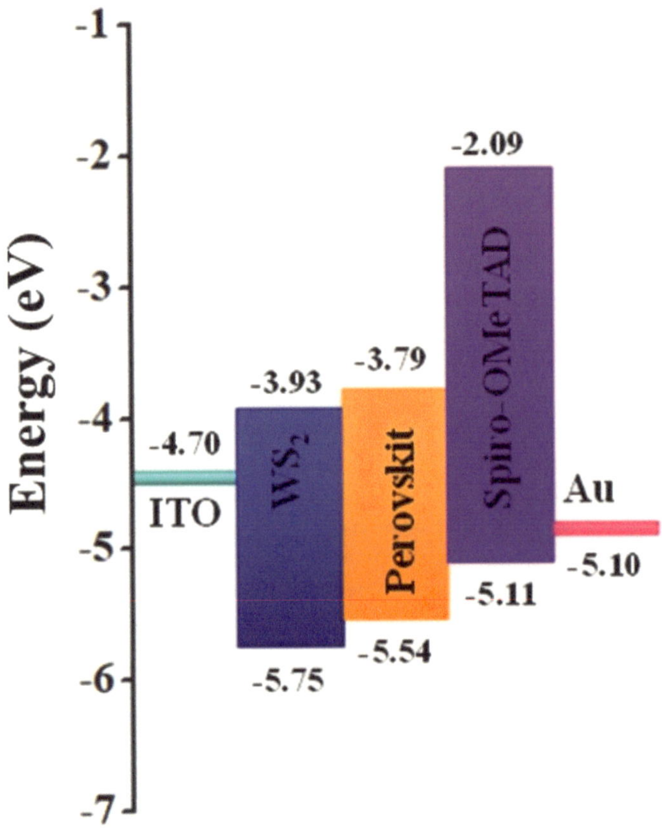

Figure 10.
Energy level diagram for n-i-p perovskite solar cells using WS_2 ETL.

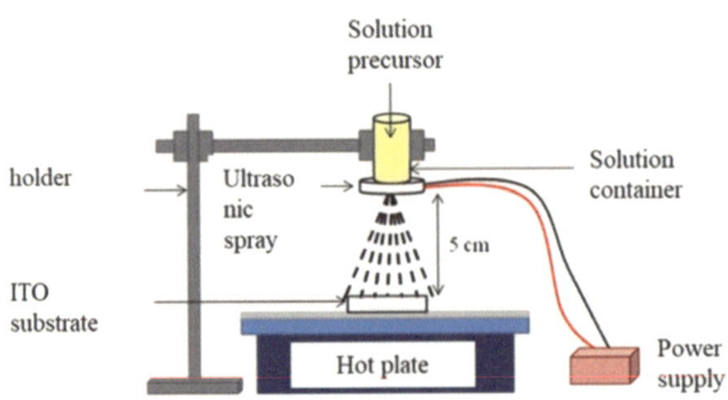

Figure 11.
Schematic diagram of ultrasonic spray pyrolysis for the preparation of TMD ETL.

The typical morphology of the WS_2 atomic layer on the ITO substrate is shown in **Figure 12A**. The WS_2 nanosheet's morphology resembles a circular structure that is produced from the precursors' mist that emerged from the ultrasonic spray membrane. Confocal Raman imaging further indicated the existence of a very thin layer of structure from the circular structure as shown in **Figure 12B**. Raman analysis then confirmed the phase crystallinity of the WS_2 (**Figure 12C**). As the figure reveals, there are two sharp peaks obtained from the Raman spectrum that is centered at 348.9 cm^{-1}

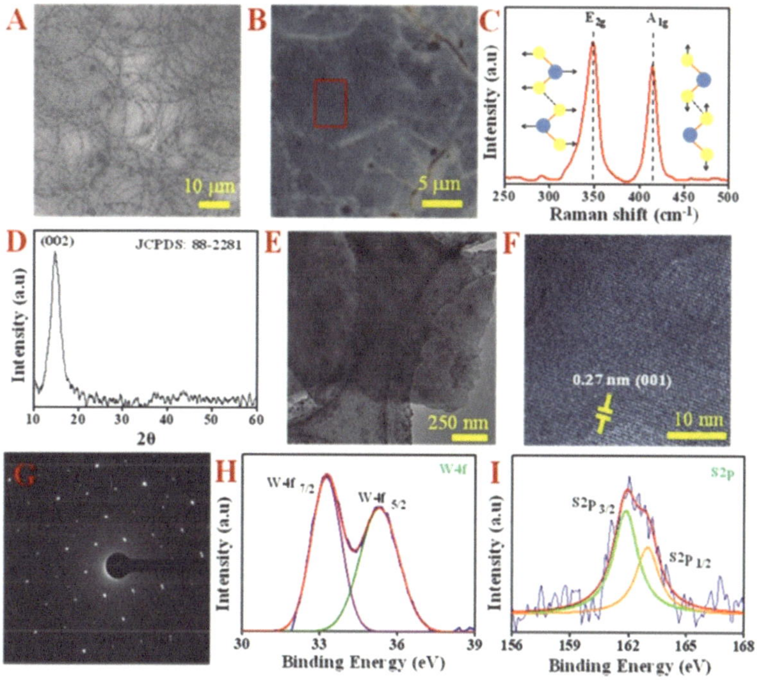

Figure 12.
The morphology, phase crystallinity, chemical state properties of WS_2 nanosheet. (A) FESEM image of WS_2 nanosheet on the ITO substrate. (B-C) Raman imaging and spectrum of WS_2 were obtained using 532 nm laser excitation. The inset in (C) shows the corresponding main vibration mode of Raman. (D) XRD spectrum for WS_2 nanosheet showing 2H phase. (E-F) Low and high-resolution TEM image of WS_2 nanosheet. (G) SAED pattern of WS_2 nanosheet showing at least three stacking WS_2 nanosheets. (H-I) High-resolution scan of XPS at W and S binding energy of WS_2 nanosheet. (Reprinted from [5]. © 2020 Wiley-VCH GmbH).

and 412.3 cm^{-1}, which are associated with the in-plane (E_{2g}) and the out-of-plane (A_{1g}) vibration modes of the lattice (see inset in **Figure 12C**) [33–39]. According to the value of the separation between these two peaks, the thickness of the atomic layer is estimated to be in the range of 10 L. The X-ray diffraction analysis further confirmed the phase crystallinity of the WS_2 layer (**Figure 12D**) [40–42]. The high-resolution transmission electron microscopy (HRTEM) and selected area electron diffraction (SAED) analysis results (**Figure 12F** and **G**) show that the sample is single crystalline. However, the presence of SAED composed of a triple spot is related to the stacking of the WS_2 atomic layer during the transfer to the lacey grid for HRTEM analysis. The XPS analysis then further confirmed the Raman and XRD analysis results on the phase crystallinity of the sample of which it belongs to WS_2 (**Figure 12H-I**).

PSCs device was fabricated using the WS_2 atomic layer as ETL and investigated how the thickness of the WS_2 ETL influenced the photovoltaic process. The structure of the PSC device is ITO/WS_2 nanosheets/Perovskite/Spiro-OMeTAD/Au. Perovskite used was triple cations system of $Cs_{0.05}[MA_{0.13}FA_{0.87}]_{0.95}Pb\ (I_{0.87}Br_{0.13})_3$ [43].

It was found that the thickness, represented by the number of layers, of the WS_2 atomic layer ETL, strongly influences the power conversion efficiency of the PSC device (**Figure 13**). The results show that the PCE performance improves with the increase of thickness from 4 L to the optimum thickness of 7 L (WS30 sample in the figure). The optimized WS_2 ETL thickness can produce a PSC device with PCE as high as 18.21% with J_{sc}, V_{oc}, and FF as high as 22.24 mA cm^{-2}, 1.12 V, and 0.731,

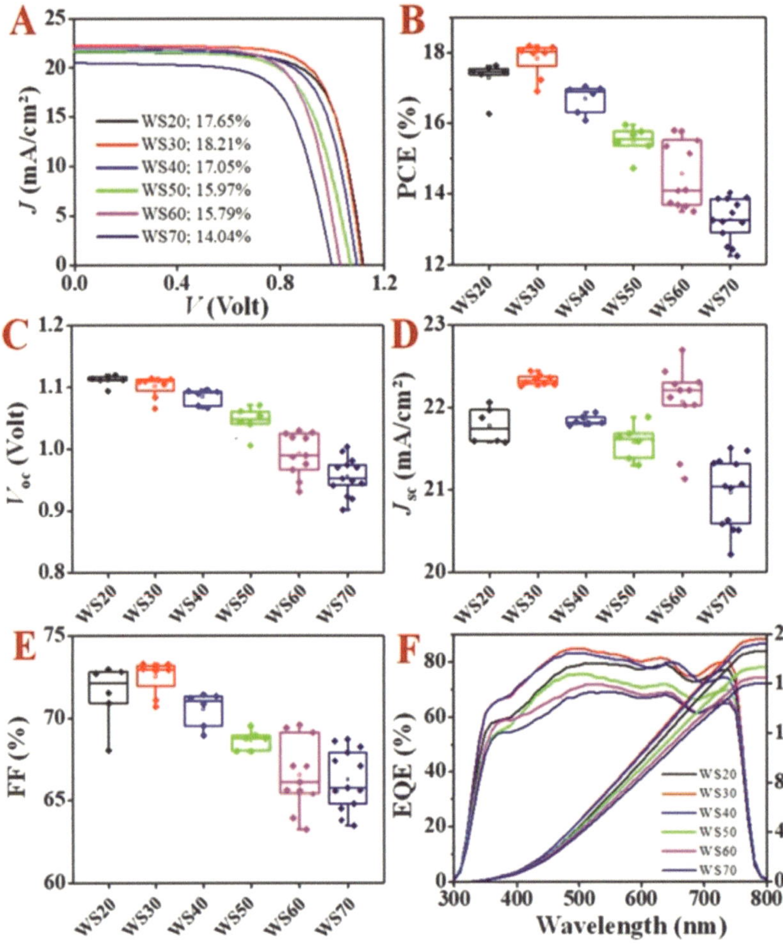

Figure 13.
The photovoltaic performance of PSC using different thicknesses of WS_2 ETL. (A) J-V curves for the champion PSC device. (B–E) The statistic plot for PCE, V_{oc}, J_{sc}, and FF, respectively. (F) EQE and integrated current density (J) of the corresponding device. (Reprinted from [5]. © 2020 Wiley-VCH GmbH).

respectively. The average performance was 17.84%, 22.33 mA cm^{-2}, 1.10 V, 0.731 for PCE, J_{sc}, V_{oc}, and FF, respectively. However, due to an increase in the energetic disorder when using the WS_2 ETL, the device performance then declined when the thickness of the ETL increased above 7 L. From **Figure 13**, we can also see that the values of V_{oc} and fill factor (FF) are impressively high, which is higher than 1.1 V for V_{oc} and approximately 74% for FF. This reflects that the photogenerated carrier dynamic in the device is high and the photogenerated carrier is effectively extracted to the external circuit to produce photocurrent [44]. This is verified by the high-external quantum efficiency (EQE) of the device as shown in **Figure 13F**.

To understand the extent effect of the WS_2 atomic layer as ETL in the PSC device, the device performance was compared with the reference PSC utilizing well-known SnO_2 ETL. In the typical process, the performance of SnO_2-based PSC shows lower performance than the WS_2 atomic layer–based device (**Figure 14**). Steady-state and transient photoluminescence analysis revealed that the interfacial charge transfer from the perovskite to ETL is high in the WS_2 atomic layer [45], the result of

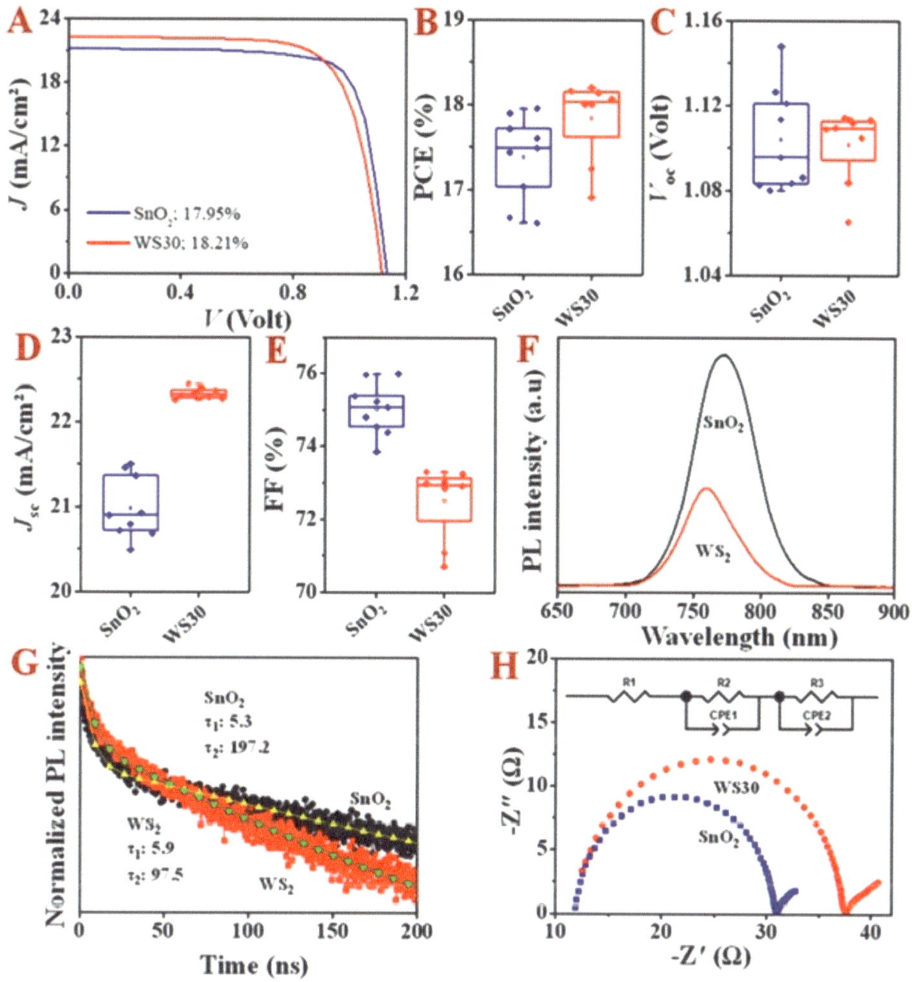

Figure 14.
The comparison of the photovoltaic parameter between WS_2 (7 L, WS30 sample) and SnO_2-based PSC device. (A) J-V curves for the champion device. (B–E) The comparison of PCE, V_{oc}, J_{sc}, and FF for the two devices, respectively. (F–H) Steady-state PL, time-resolved PL spectra (TRPL), and electrochemical impedance spectra for WS_2 and SnO_2-based PSC devices, respectively. (Reprinted from [5]. © 2020 Wiley-VCH GmbH).

optimized coupling due to ultra-flat surface morphology offered by the WS_2 atomic layer. This phenomenon is further confirmed by the electrochemical impedance spectroscopy analysis result where it is obtained that the interface charge transfer resistance is lower in the WS_2-based PSC device than the SnO_2-based device. Thus, it can be remarked that the WS_2 atomic layer enables highly active interfacial charge transfer for a high-performance PSC device.

3. Conclusions

2D atom thick TMD promises facile charge extraction and transport in the perovskite solar cells due to its ultimate thin and single-crystalline nature. The optimization of the 2D TMD layer to obtain a large dimension on the substrate surface is necessary to further promote a highly dynamic photogenerated carrier in the perovskite solar cells device. These materials may become a potential platform for high-performance perovskite solar cells.

Acknowledgements

We acknowledged the financial support from the Universiti Kebangsaan Malaysia for supporting this project under GUP-2019-071 and DIP-2021-025."

Conflict of interest

The authors declare no conflict of interest.

Author details

Akrajas Ali Umar*, Nurul Ain Abd Malek, Nabilah Alias and Abang Anuar Ehsan
Institute of Microengineering and Nanoelectronics, Universiti Kebangsaan Malaysia, Selangor, Malaysia

*Address all correspondence to: akrajas@ukm.edu.my

References

[1] Jeong J, Kim M, Seo J, Lu H, Ahlawat P, Mishra A, et al. Pseudo-halide anion engineering for α-FAPbI 3 perovskite solar cells. Nature. 2021; **592**(7854):381-385

[2] Lin PY, Loganathan A, Raifuku I, Li MH, Chiu YY, Chang ST, et al. Pseudo-Halide Perovskite Solar Cells. Advanced Energy Materials. 2021;**2100818**

[3] Deng X, Cao Z, Yuan Y, Chee MOL, Xie L, Wang A, et al. Coordination modulated crystallization and defect passivation in high quality perovskite film for efficient solar cells. Coordination Chemistry Reviews. 2020;**420**:213408

[4] Jeong I, Park YH, Bae S, Park M, Jeong H, Lee P, et al. Solution-Processed Ultrathin TiO2 Compact Layer Hybridized with Mesoporous TiO2 for High-Performance Perovskite Solar Cells. ACS Applied Materials & Interfaces. 2017;**9**(42):36865-36874

[5] Malek NAA, Alias N, Umar AA, Zhang X, Li X, Saad SKM, et al. Enhanced Charge Transfer in Atom-Thick 2H–WS2 Nanosheets' Electron Transport Layers of Perovskite Solar Cells. Solar RRL. 2020;**4**(10):2000260

[6] Umar AA, Zhang X, Saad SKM, Abd Malek NA, Liu K, Alias N, et al. Enhancing the interfacial carrier dynamic in perovskite solar cells with an ultra-thin single-crystalline nanograss-like TiO 2 electron transport layer. Journal of Materials Chemistry A. 2020;**8**(27):13820-13831

[7] Huang P, Yuan L, Zhang K, Chen Q, Zhou Y, Song B, et al. Room-temperature and aqueous solution-processed two-dimensional TiS2 as an electron transport layer for highly efficient and stable planar n–i–p perovskite solar cells. ACS Applied Materials & Interfaces. 2018;**10**(17):14796-14802

[8] Sumpter BG, Liang L, Nicolai A, Meunier V. Interfacial properties and design of functional energy materials. Accounts of Chemical Research. 2014;**47**(11):3395-3405

[9] Kuc A, Heine T. On the Stability and Electronic Structure of Transition-Metal Dichalcogenide Monolayer Alloys Mo1− xXxS2− ySey with X= W, Nb. Electronics. 2016;**5**(1):1

[10] Furchi MM, Pospischil A, Libisch F, Burgdörfer J, Mueller T. Photovoltaic effect in an electrically tunable van der Waals heterojunction. Nano Letters. 2014;**14**(8):4785-4791

[11] Van Le Q, Choi J-Y, Kim SY. Recent advances in the application of two-dimensional materials as charge transport layers in organic and perovskite solar cells. FlatChem. 2017;**2**:54-66

[12] Ren Y, Yang X, Zhou L, Mao JY, Han ST, Zhou Y. Recent advances in ambipolar transistors for functional applications. Advanced Functional Materials. 2019;**29**(40):1902105

[13] Singh R, Giri A, Pal M, 7Thiyagarajan K, Kwak J, Lee J-J, et al. Perovskite solar cells with an MoS 2 electron transport layer. Journal of Materials Chemistry A. 2019;**7**(12):7151-7158

[14] Yin G, Zhao H, Feng J, Sun J, Yan J, Liu Z, et al. Low-temperature and facile solution-processed two-dimensional TiS2 as an effective electron transport layer for UV-stable planar perovskite solar cells. Journal of Materials Chemistry A. 2018;**6**(19):9132-9138

[15] Singh R, Giri A, Pal M, Thiyagarajan K, Kwak J, Lee J-J, et al. Perovskite solar cells with an MoS2 electron transport layer. Journal of Materials Chemistry A. 2019;7(12):7151-7158

[16] Abd Malek NA, Alias N, Saad SKM, Abdullah NA, Zhang X, Li X, et al. Ultra-thin MoS2 nanosheet for electron transport layer of perovskite solar cells. Optical Materials. 2020;**104**:109933

[17] Ahmed MI, Hussain Z, Khalid A, Amin HMN, Habib A. Absorption enhancement in CH3NH3PbI3solar cell using a TiO2/MoS2nanocomposite electron selective contact. Materials Research Express. 2016;**3**(4):045022

[18] Huang P, Chen Q, Zhang K, Yuan L, Zhou Y, Song B, et al. 21.7% efficiency achieved in planar n–i–p perovskite solar cells via interface engineering with water-soluble 2D TiS 2. Journal of Materials Chemistry A. 2019;7(11):6213-6219

[19] Jung DH, Oh YJ, Nam YS, Lee H. Effect of layer number on the properties of stable and flexible perovskite solar cells using two dimensional material. Journal of Alloys and Compounds. 2021;**850**:156752

[20] Dasgupta U, Chatterjee S, Pal AJ. Thin-film formation of 2D MoS2 and its application as a hole-transport layer in planar perovskite solar cells. Solar Energy Materials & Solar Cells. 2017;**172**:353-360

[21] Zhao E, Gao L, Yang S, Wang L, Cao J, Ma T. In situ fabrication of 2D SnS2 nanosheets as a new electron transport layer for perovskite solar cells. Nano Research. 2018;**11**(11):5913-5923

[22] Chu W, Li X, Li S, Hou J, Jiang Q, Yang J. High-Performance Flexible Perovskite Solar Cells with a Metal Sulfide Electron Transport Layer of SnS2 by Room-Temperature Vacuum Deposition. ACS Applied Energy Materials. 2019;**2**(1):382-388

[23] Sobayel K, Akhtaruzzaman M, Rahman K, Ferdaous M, Al-Mutairi ZA, Alharbi HF, et al. A comprehensive defect study of tungsten disulfide (WS2) as electron transport layer in perovskite solar cells by numerical simulation. Results in Physics. 2019;**12**:1097-1103

[24] Gao J, Gupta M. Titanium disulfide as Schottky/ohmic contact for monolayer molybdenum disulfide. npj 2D Materials and Applications. 2020;**4**(1):26

[25] Alias N, Ali Umar A, Malek NAA, Liu K, Li X, Abdullah NA, et al. Photoelectrical Dynamics Uplift in Perovskite Solar Cells by Atoms Thick 2D TiS2 Layer Passivation of TiO2 Nanograss Electron Transport Layer. ACS Applied Materials & Interfaces. 2021;**13**(2):3051-3061

[26] Capasso A, Matteocci F, Najafi L, Prato M, Buha J, Cinà L, et al. Few-Layer MoS2 Flakes as Active Buffer Layer for Stable Perovskite Solar Cells. Advanced Energy Materials. 2016;**6**(16):1600920

[27] Ambrosi A, Sofer Z, Pumera M. 2H --> 1T phase transition and hydrogen evolution activity of MoS2, MoSe2, WS2 and WSe2 strongly depends on the MX2 composition. Chemical Communications (Cambridge). 2015;**51**(40):8450-8453

[28] Xiao S, Xiao P, Zhang X, Yan D, Gu X, Qin F, et al. Atomic-layer soft plasma etching of MoS2. Scientific Reports. 2016;**6**:19945

[29] Wang L, Hu P, Long Y, Liu Z, He X. Recent advances in ternary two-dimensional materials: synthesis, properties and applications. Journal of Materials Chemistry A. 2017;**5**(44):22855-22876

[30] Jeon J, Jang SK, Jeon SM, Yoo G, Jang YH, Park JH, et al. Layer-controlled CVD growth of large-area two-dimensional MoS2 films. Nanoscale. 2015;7(5):1688-1695

[31] Ahmed MI, Hussain Z, Khalid A, Amin HMN, Habib A. Absorption enhancement in CH3NH3PbI3 solar cell using a TiO2/MoS2 nanocomposite electron selective contact. Materials Research Express. 2016;**3**(1):015022

[32] Park J, Woo H, Jeon S. Impact of fast transient charging and ambient on mobility of WS2 field-effect transistor. Journal of Vacuum Science & Technology, B: Nanotechnology & Microelectronics: Materials, Processing, Measurement, & Phenomena. 2017;**35**(5):050601

[33] Gutiérrez HR, Perea-López N, Elías AL, Berkdemir A, Wang B, Lv R, et al. Extraordinary Room-Temperature Photoluminescence in Triangular WS2 Monolayers. Nano Letters. 2013;**13**(8):3447-3454

[34] Qiao S, Yang H, Bai Z, Peng G, Zhang X. Identifying the number of WS2 layers via Raman and photoluminescence spectrum. Advances in Engineering Research. 2017;**141**:1408-1413

[35] Park J, Kim MS, Cha E, Kim J, Choi W. Synthesis of uniform single layer WS2 for tunable photoluminescence. Scientific Reports. 2017;**7**(1):16121

[36] Zhang Y, Shi J, Han G-F, Li M, Ji Q, Ma D, et al. Chemical vapor deposition of monolayer WS2 nanosheets on Au foils towards direct application in hydrogen evolution. Nano Research. 2015;**8**:2881-2890

[37] Molas MR, Nogajewski K, Potemski M, Babinski A. Raman scattering excitation spectroscopy of monolayer WS2. Scientific Reports. 2017;**7**(1):5036

[38] Zhao W, Ghorannevis Z, Amara KK, Pang JR, Toh M, Zhang X, et al. Lattice dynamics in mono- and few-layer sheets of WS2 and WSe2. Nanoscale. 2013;**5**:9677-9683

[39] Berkdemir A, Gutiérrez HR, Botello-Méndez AR, Perea-López N, Elías AL, Chia C-I, et al. Identification of individual and few layers of WS2 using Raman Spectroscopy. Scientific Reports. 2013;**3**:1755

[40] Jian-Ping Z, Jun M, Jin-Ming L, Jian Y, Junkai H, Yongtao M, et al. Fabrication of novel heterostructured few layered WS 2-Bi 2 WO 6/Bi 3.84 W 0.16 O 6.24 composites with enhanced photocatalytic performance. Applied Catalysis B: Environmental. 2015;**179**:220-228

[41] Sang Y, Zhao Z, Zhao M, Hao P, Leng Y, Liu H. From UV to near-infrared, WS2 nanosheet: a novel photocatalyst for full solar light spectrum photodegradation. Advanced Materials. 2015;**27**(2):363-369

[42] Hai X, Chang K, Pang H, Li M, Li P, Liu H, et al. Engineering the Edges of MoS2 (WS2) Crystals for Direct Exfoliation into Monolayers in Polar Micromolecular Solvents. Journal of the American Chemical Society. 2016;**138**(45):14962-14969

[43] Saliba M, Matsui T, Seo JY, Domanski K, Correa-Baena JP, Nazeeruddin MK, et al. Cesium-containing triple cation perovskite solar cells: improved stability, reproducibility and high efficiency. Energy & Environmental Science. 2016;**9**(6):1989-1997

[44] Kim G-W, Shinde DV, Park T. Thickness of the hole transport layer in perovskite solar cells: performance versus reproducibility. RSC Advances. 2015;**5**(120):99356-99360

[45] Wang F, Chen Y, Han G, Zhang Q, Ma Q. The interface and its role in carrier transfer/recombination dynamics for the planar perovskite solar cells prepared under fully open air conditions. Current Applied Physics. 2016;**16**(10):1353-1363

Chapter 6

Advanced Chalcogen Cathode Materials for Lithium-Ion Batteries

Varishetty Madhu Mohan, Madhavi Jonnalagadda and VishnuBhotla Prasad

Abstract

As on today the main power sources of lithium-ion batteries (LIBs) research developments gradually approach their theoretical limits in terms of energy density. Therefore, an alternative next-generation of power sources is required with high-energy densities, low cost, and environmental safety. Alternatively, the chalcogen materials such as sulfur, selenium, and tellurium (SSTs) are used due to their excellent theoretical capacities, low cost, and no toxicity. However, there will be some challenges to overcome such as sluggish reaction of kinetics, inferior cycling stability, poor conductivity of S, and "shuttle effect" of lithium polysulfides in the Li-S batteries. Hence, several strategies have been discussed in this chapter. First, the Al-SSTs systems with more advanced techniques are systematically investigated. An advanced separators or electrolytes are prepared with the nano-metal sulfide materials to reduce the resistance in interfaces. Layered structured cathodes made with chalcogen ligand (sulfur), polysulfide species, selenium- and tellurium-substituted polysulfides, $Se_{1-x}S_x$ uniformly dispersed in 3D porous carbon matrix were discussed. The construction of nanoreactors for high-energy density batteries are discussed. Finally, the detailed classification of flexible sulfur, selenium, and tellurium cathodes based on carbonaceous (e.g., carbon nanotubes, graphene, and carbonized polymers) and their composite (polymers and inorganics) materials are explained.

Keywords: sulfur, selenium, tellurium (SSTs) electrodes, polysulfides, nano-metal chalcogen and flexible batteries, nanoreactors, electrochemical properties

1. Introduction

Recently, much attention has been focused on the development of more safe, high-energy density, long-life, and low-cost batteries to satisfy our energy demand as our daily life includes electric vehicles, portable electronics, and large-scale grids [1–4]. However, lithium-ion batteries (LIBs) successfully prepared and available in the commercial market since the 1990s, even though their theoretical specific capacity, energy density of the electrode material is relatively low. Hence, still it is challenging to development of next generation of lithium-ion batteries to fulfill the demand by overcoming their hindrance by the intrinsic limitation [5].

Various metal sulfides, metal oxides, and metal poly-anions have been developed, and they exhibited preferable capacities, but their output voltages are not sufficient. Hence, chalcogen materials (sulfur, selenium, and tellurium (SSTs) are emerging conversion-type cathode materials for aluminum-ion batteries (AIBs) [6]. They exhibit the multi-electron transfer process; therefore, AlSST batteries can deliver very high capacities.

The low-cost and -toxicity, most-abundant sulfur (S) has theoretical gravimetric capacity of 1672 mA h g^{-1} due to the reaction between S and sulfide (S^{2-}) exhibiting the highest capacity, over all cathode materials in AIBs [7]. Notably, lithium-sulfur (Li-S) batteries exhibited the highest theoretical energy density of 2600 Wh kg^{-1}. Moreover, the sulfur-based cathodes require a large number of conductive additives due to their electrical insulating property to avoid leakage/loss issues, which greatly decreases the actual capacity [8]. Therefore, it is usually difficult to achieve the theoretical specific capacity. Based on these merits, metal chalcogens batteries (MCBs) are an attractive interest for alternative of new-generation secondary batteries.

The main research has been focused on lithium dendrite and "shuttle effect" of high-order lithium phosphate sulfur batteries. There is three-phase boundary between cathodes, electrolytes, and modified interface layers. The interface boundaries occupied by the sulfur species restrict the redox kinetics. Therefore, to improve the redox reaction, it is necessary to make strong bond of S to the host, thereby continuing to conduct mixed charge carriers (Li^+ and e^-). There is another difficulty that the degradation of active materials between the electrodes due to shuttle process, in which cathode electrolyte interfaces bring rapid decay of the capacity, thereby reducing the coulomb efficiency of Li-S batteries [2]. The formation of lithium dendrites on the surface of lithium anode and also unstable solid electrolyte interface (SEI) leads to low columbic efficiency (CE) and poor cyclic performance [9]. To overcome these problems, most suitable nanomaterials are considered for energy applications due their unique crystal structure providing high surface-area-to-volume ratio and shortening lithium ions transport [9].

In order to create the metal-ligand covalency is the one of strategies by replacing the oxide ligand with the chalcogen (S, Se) to achieve an anion redox stabilization, where the less electronegative nature of the chalcogen improves the ligand p band penetration into the metal d band. Tarascon et al. investigated layered chalcogen structures as well as their electrochemical performance for the next generation of cathodes [10]. The results exhibited the superior performance in voltage and capacity fade with voltage hysteresis. Hence, chalcogen anion redox plays a critical role in a Li-rich cathode batteries. Some of research was carried out on chalcogen cathodes, both Li-rich and conventional chalcogen cathodes for the evolution of chalcogen anion redox cathode [11, 12]. Mespoulie et al. [13] introduced fast Li-ion conductors of mixed anionic and cationic redox activity of Li_2SnS_3, by introducing the Fe redox couple in the host cation Sn site.

Selenium (Se) is another class of material present in Al-Se battery that showed higher voltage plateau resulting a desirable energy density [14]. Selenium metal has high theoretical volumetric capacity (3253 mA h cm^{-3}, $\rho = 4.81$ g cm^{-3}), which is more suitable especially in hybrid electric vehicles and in the mobile smart phones due to restrictions of the battery volume [15]. Selenium showed higher electronic conductivity (1×10^{-3} S m^{-1}) and excellent kinetic behavior than sulfur [16]. The chemical compound $Se_{1-x}S_x$ with different Se-S ratios shows higher theoretical capacity as well as better electronic conductivity due to fast reaction of kinetics than pristine S [17]. Even though, Se_{1-x} S_x cathode materials also suffer from poor cycle performance, lower coulombic efficiency due to the dissolution and shuttling of intermediates [18]. The electrochemical performance of Li-$Se_{1-x}S_x$ batteries improved by carbon coating, which provide a strong chemical affinity of polarized surface, which can effectively trap the soluble intermediates to minimize the shuttle effect and side reactions in the electrolyte [19]. A series of $Se_{1-x}S_x$ cathodes were prepared by Se/S ratio and the presence of supercritical CO_2. NC@SWCNTs@$Se_{1-x}S_x$ cathodes exhibited higher conductivity and strong adsorption leading to superior cyclic efficiency.

Tellurium (Te) material has the highest atomic weight among sulfur and selenium, high electrical conductivity and a 6-electron transfer reaction process made it to be promising cathode material in AIBs [20]. However, still there are several challenges remaining to overcome for the development of batteries such as low electronic conductivity of S, shuttle effect, slow kinetics of ionic liquids as well as undesirable reaction mechanism [21]. Tellurium exhibits a higher theoretical volumetric capacity of 2619 mAh cm^{-3} due to its intrinsic electrical conductivity of Te (2×10^2 S m^{-1}), much better than that of S (5×10^{-16} S m^{-1}) and Se (1×10^{-4} S m^{-1}). Therefore, the high utilization ratio of active material of Te leads to good performance at the large current density. The fabricated batteries based on Te/porous carbon (Te-G-CNT) electrode materials deliver a high volumetric capacity up to 2493.13 mAh cm^{-3} [22].

The polysulfide (Li_2Sn) species have strong tendency to catenate and form reactive polysulfide dianions as well as radical anions (Sn^{2-} and $Sn/_{n/2}^{-}$, $2 < n < 8$). These conversion reactions of sulfur ↔ Li_2S kinetically favored in the mediated solution and their deposition degrading the lithium surface and the cyclic stability [23–25]. The dissolved species shows shuttle effect by insulating deposition of Li_2S/Li_2S_2. Polysulfide molecules modified by substituting chalcogen atoms minimized the intrinsic shuttle effect [26]. By substituting S, Se and Te can be facilely formed as the polyselenosulfides ($Li_2Se_xS_y$) and polytellurosulfides ($Li_2Te_xS_y$). However, selenium and tellurium lead to significant differences in the electrochemical performance compared with Li-S batteries. The substitution of selenium and tellurium has significant impact on the metal-chalcogen batteries and solid-state batteries by employing chalcogenide solid electrolytes. Therefore, in this chapter, the strategies to improve electrochemical performance are elaborated, and the development of new trends for next-generation lithium-ion batteries is provided.

To fabricate flexible lithium-ion batteries using sulfur-based cathodes, there are two main synthetic approaches: (1) Post-sulfur loading: The formation of a flexible skeleton then loaded with sulfur by using vapor infusion, melt diffusion, or reprecipitation of sulfur from a solution (generally carbon disulfide (CS_2) or toluene). (2) Pre-sulfur loading: pre-synthesized sulfur composites into a flexible cathode. By keeping the flexible cathode required features in mind such as (1) high content of active materials with respect to total mass of the electrode, (2) mechanically robust skeleton, (3) long-range interpenetrated conductive network, (4) porous structure, and (5) three-dimensional (3D) scaffold to improve areal sulfur loading [27]. The flexible energy Li-S batteries, flexible alkali metal-chalcogen batteries, and two special flexible batteries such as prototypes of foldable and cable-type Li-S batteries are discussed.

2. Results and discussion of various topics

2.1 Sulfur, selenium and tellurium batteries

Many efforts have focused on the cathode material design, electrolyte optimizations, separator modification, still some of the challenges remain due to slow kinetics, electrolyte compatibility, and inferior cycling stability. Hence, there are many possibilities for the development of more reliable sulfur, selenium tellurium (SSTs) batteries. He et al. [5] showed schematic representation of various components for the development of lithium-ion batteries based on SST as shown in **Figure 1**. The carbon-based materials, conductive materials, and their nanostructure with a porous matrix refereed as a host due to low electrical conductivity of SSTs (S and Se) and the soluble properties of the chalcogenide. Therefore, the host materials can provide necessary contact with SSTs to reduce the formation of inactive regions and satisfy the adsorbing as well as accommodation of soluble active materials. Therefore, the mass loading of SSTs cathodes can efficiently be increased. The various approaches such as melting diffusion, chalcogen vaporization are used to increase the mass loading of active SSTs into the conductive host materials. An introduced of metal atoms to form a bonding with SSTs is another possible method for reducing the reaction barriers in Al-SSTs batteries.

2.1.1 Al-sulfur batteries

The most abundant high-surface-area carbon porous materials are possible to absorb the SSTs materials, which is more impartment for limiting chalcogenide dissolution [28]. Thus, well-designed porous structure carbon materials composites can enhance not only the charge transport but also improve the retention of SSTs cathode during electrochemical reaction [29]. Therefore, carbon materials are attractive to be host for the insulating S with a regular matrix. A melt-diffusion method was conducted to prepare the S/activated carbon cloth (ACC) composite cathode material in Al-S battery as shown in **Figure 1a**. As prepared ACC material exhibited type I adsorption, corresponding to microporous structure with a pore size below 2 nm. The Brunnauer-Emmett-Teller (BET) measurement showed that after compositing of S into the ACC, specific surface area decreased from 2376.6 to 1532.8 m^2 g^{-1} and also decreased to its micropore volume from 0.93 to 0.61 cm3 g^{-1}, indicating that the S material was uniformly impregnate into the microporous structure. The Al-S battery based on S/ACC cathode exhibited a high specific capacity of 1320 mA h g^{-1} and discharged voltage about 0.65 V as shown in **Figure 2b**.

The microporous structure ACC host consists of pore size less than 2 nm and can effectively provide the fast solid-state reaction kinetics favoring to its ready electron

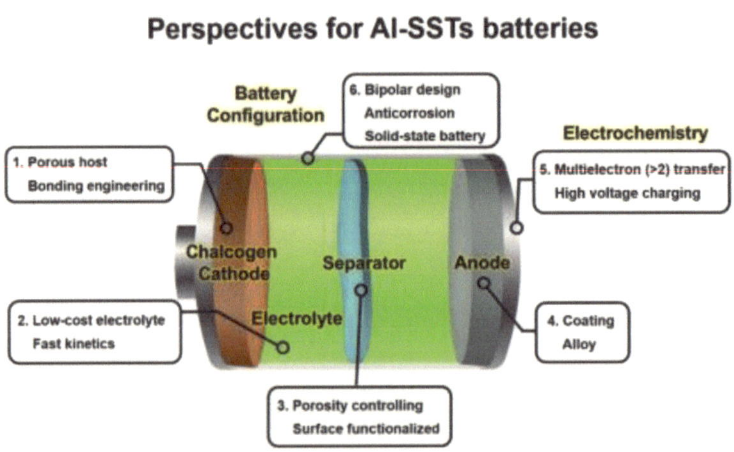

Figure 1.
Perspective for Al-SSTs batteries. Reprinted from ref. [5].

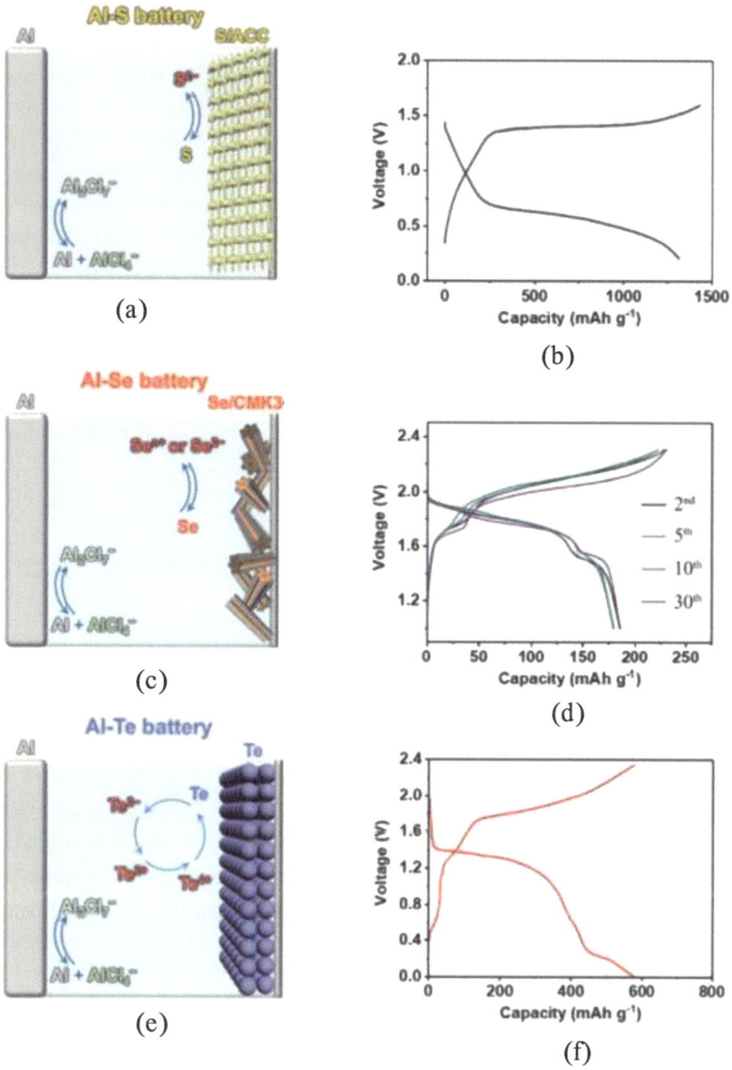

Figure 2.
The galvanostatic charge/discharge curves of a, b) Al-S battery, c, d) Al-Se battery (n ≥ 1), and e, f) Al-Te battery. Reprinted from ref. [5].

access, large reaction area, and decreasing the ionic diffusion length. Similarly, a free-standing CNF host was also introduced in Al-S battery as shown in 2b. The carbon nanofiber with a diameter of 100–200 nm occupies between the interspaces in micrometer scale level. This CNF structure provides a spacious, robust, conductive matrix to accommodate the active S and their products. The S and EMIC/$AlCl_3$ slurry dispersed into the freestanding CNF host, the Al-S battery exhibited a good capacity of 1250 $mAhg^{-1}$. These above free-standing carbon materials not only provide a conductive matrix for S materials, but also reduce the side reaction from the binder, thereby enhanced stability of Al-S battery. The porous carbonized Cu-based metal organic frameworks (MOFs) called as HKUST-1-C also introduced as a host to the S in Al-S batteries [30]. The HKUST-1-C carbon materials exhibited high hierarchical porous structure with surface area of 179 $m^2\ g^{-1}$ and a pore sizes in <5 nm range.

These are more suitable for being host in S cathode batteries. The metallic Cu can react with polysulfides to form S-Cu ionic clusters, thereby reducing the kinetic barrier of the electrochemical conversion reaction and facilitating the reversibility of S during charge/discharge processes. Therefore, the S@HKUST-1-C cathode battery exhibited a stable performance with a reversible capacity of 600 mA h g^{-1} at the 75th cycle and retained 460 mA h g^{-1} even after 500 cycles at 1 A g^{-1}. These results indicate that the metallic material provides a valuable strategy to develop stable Al-S batteries. A nitrogen-doped hierarchical porous carbon called as HPCK used as a host for S in Al-S batteries [31]. Hierarchical micro-, meso-, and macro-pores of HPCK was prepared by carbonizing a N-rich polymer precursor combined with zinc nitrate and followed by a KOH etching process. Interestingly, KOH etching process greatly improved the surface area (2513 m^2 g^{-1}) and created more micro- and meso-pores. The Al-S batteries based on S/HPCK cathode delivered a capacity of 1027 mA h g^{-1} at 0.2 A g^{-1} for 50 cycles and exhibited excellent cyclic ability of 405 mA h g^{-1} at 1 A g^{-1} for 700 cycles. The hierarchical porous structure of HPCK with high surface area and large pores confined S materials. The huge macropores provide fast ion transport in the electrolyte. Therefore, the porous carbon powders, metal content along with structure optimization for host are impartment factors for Al-S batteries.

2.1.2 Al-selenium batteries

Selenium composite prepared with carbon meso-porous material (2 nm < pore size <50 nm) can enhance the encapsulation of Se in the Al-Se batteries as shown in **Figure 2C** [32]. The CMK-3 carbon demonstrated that it exhibited a hexagonal meso-porous structure favorable for being host material in Al-Se batteries as shown in **Figure 2c**. The CMK-3 carbon nanorods showed a large pore volume of 1.78 cm^3 g^{-1}, a high surface area about 1632 m^2 g^{-1}, and a uniform pore size of 3.4 nm. The Al-Se battery based on Se@CMK-3 cathode delivers an initial discharge capacity of 218 mA h g^{-1} at 100 mA g^{-1} and a relatively high discharged potential above 1.5 V as shown in **Figure 2d**. Hollow selenium carbon nanotube (Se@CT) with a specific surface area of 61.49 m^2 g^{-1} and pore diameter of 3.36 nm was also demonstrated as cathode material in Al-Se batteries [6]. The Se@CT cathode exhibited an initial discharge capacity of 447.2 mA h g^{-1} at 200 mA g^{-1} with a voltage of about 1.6 V. The capacity retains 83.5% even after the 200 cycles at 500 mA g^{-1}. The carbon materials in Se@ CT cathode reside their structural stability of Se, reduce the dissolution of selenide products, and also avoid the volume change of Se during the electrochemical process. Therefore, mesoporous structure and pore size are important to stabilize the Se cathode as well as enhance the electrochemical performance. Mesoporous carbon fibers (MCFs) size from 2.7 to 8.9 nm prepared and demonstrated the effect of pore size on the electrochemical performance of Al-Se batteries [32]. The Al-Se battery based on MCFs material with pore size of 7.1 nm exhibited a good capacity of 366 mA h g^{-1}. The chloroaluminate ion diffusivity greatly affects in the mesopore size of MCFs composite Se cathode. The carbon well-designed structures may provide the chloroaluminate ion transportation as well as charge transportation during charge/discharge processes, which enhances the electrochemical behavior of Al-Se batteries.

2.1.3 Al-Te cathode batteries

The electrical conductivity of Te cathode is very high [12]. Hence, utilization ratio of active material is large and also exhibited good rate performance in Al ion batteries.

Te cathode is easily prepared by coating the slurry of Te powder, acetylene black, and binder on current collector, without any host materials as shown in **Figure 2e** [20]. The Al-Te battery exhibited with raw material an initial capacity of 913 mA h g^{-1} at 20 mA g^{-1} with a potential of 1.4 V as shown in **Figure 2f**. The Al-Te battery delivers a good rate performance at different current densities due to the high electrical conductivity of Te. However, Te batteries exhibited capacity fading due to its leaching of soluble telluride from the cathode. Further, the rGO materials were introduced to encapsulate Te nanowires in Al-Te battery [6]. This Te/rGO cathode battery exhibited a capacity of 1026 mA h g^{-1} at 500 mA g^{-1} and also delivers a considerable capacity beyond 100 cycles at 1.0 and 2.0 A g^{-1}. The rGO materials suppress the dissolution of telluride into electrolyte indicating better utilization of Te. Further, N-doped porous carbon materials coupled with rGO also introduced as a host materials for the improvement of the stability of Al-Te batteries [33]. The rGO materials are easily encapsulated soluble tellurium species under physical and chemical confinements. Therefore, the Al-Te batteries exhibited excellent cyclic ability. It exhibited initial specific capacity of 935.5 mA h g^{-1} and 467.5 mA h g^{-1} after 150 cycles with the Te loading of 70 wt%. Thus, a host material with well-designed structure, such as porous conductive matrix with specific components is necessary for cathode materials in Al-SSTs batteries.

2.2 Interfacial problems in sulfur batteries

In the case of sulfur lithium-ion batteries, during the discharge/charge reaction, the sulfur required to be tightly attached to a host with sustainable conduction of Li^+ and e^-. Generally, the cathodic reaction occurs at the host/elemental sulfur/liquid electrolyte interface. The carbon nano materials such as graphene, carbon nanotube (CNT), or carbon nanofiber and metal sulfide are the indispensable 2D materials to sulfur host. The modified host materials with only nano-sized pores alone cannot accommodate the sulfur and completely reduce suppression of the shuttle effect in the LiPS [34]. Therefore, in recent years the development of interface components for Li-S batteries is most impartment [28, 35]. One of the main strategies is that the coating of materials exhibited several merits such as fast electrical and ionic transmission capability, uniform thickness, and stable distribution of composition on the surface of cathode materials without effect of volume.

The another most prominent strategy is that in situ growth of nano transition metal oxides (sulfide) [36] or the loading of nano transition metals [37, 38] on the carbon surface leads to overcome the poor contact between sulfur and carbon materials. In this process, (sulfides, oxides, nitrides, etc.) compounds are added or doped some of the elements (N, S and co-NS) and their derivatives. Therefore, polar bonds are generated between host and sulfur; those provided fast transmission of electrons as well as increase the ions redox reactions at the interface.

2.2.1 Inter facial problems in nano metal sulfides (oxide)

Various nano metal oxides have interacted with LiPS through strong chemical bonds, which are reducing the shuttle effect. In particular, oxygen-rich compounds such as V_2O_3 [39], TiO_2 [40, 41], SnO_2 [42], Co_3O_4 [43], MnO_2 [44, 45]) successfully prepared as LiPS traps to enhance the cyclic stability. The high conductive Ni foam/graphene/carbon nanotubes/MnO_2 nanoflakes (NGCM) were proposed in which interconnected Ni foam, graphene, and carbon nanotubes of the NGCM sponge facilitated efficient electron transfer. The NGCM sponge showed good wettability and interfacial contact with the Li-S electrolyte, and the MnO_2 nano flakes exhibited electro-catalytic effects as well as strong chemisorption on LiPS [46]. The porous and double-shelled architecture decreases the ion transfer distance, Uniform sulfur distribution offers active interfaces as well as decreases volume changes. Luo et al. developed spinel Ni-Co oxide double-shelled microspheres (NCO-HS), which consisted of defective spinel $NiCo_2O_{4-x}$, as the multifunctional sulfur host material. The S@NCO-HS prepared under high sulfur loading exhibited minimum capacity fading rate of 0.045% per cycle over 800 cycles with high areal specific capacity of 6.3 mAh cm^{-2} at 5 C.

2.2.2 Capping layer of MTO-CNTs for sulfur cathode

The thin barrier layers designed with light weight and with high polysulfide-trapping capabilities showed high weight density (usually >0.3 mg cm^{-2}), unexpectedly reducing the overall energy densities of Li-S batteries as shown in **Figure 3** [47–55]. The development of lightweight MTO-CNTs capping layer directly coated onto the surface of sulfide cathode as shown in **Figure 4a**. The MTO-CNTs capping layer prepared on the sulfur cathode, which is directly contacted with an electrically conducting layer to form a cathodic "sub cell" for capturing and decreasing the polysulfide species. The thickness (~2 μm) and the weight density (0.06 mg cm^{-2}) of the MTO-CNTs capping layer are much lower than other barrier layers reported elsewhere, as the mesoporous carbon and the grapheme layers [47, 48]. Moreover, it is noted that the area of conventional interlayer is higher than the coated capping layer. To understand the layer-by-layer electrode structure and the cathode structure was observed by SEM and energy dispersive X-ray spectroscopy (EDX). The 1D, MTO-CNTs nanostructure and the ultrathin capping layer are formed through self-weaving and firmly coated on the surface of sulfur cathode (**Figure 4b**). The existence of MTO-CNTs capping layer is estimated by the elemental mapping analysis (**Figure 4c, d**). The MTO-CNTs capping layer formed on the cathode can trap more effectively, polysulfides within the cathode material, thereby reducing the shuttle effect of polysulfide. The CV curves of electrodes reveal that the electrode with MTO-CNTs capping layer has sharper and more intensive oxidation and reduction peaks exhibited than an electrode without capping layer, indicating that the capping layer efficiently enhanced reaction kinetics of the electrode. The substantially promoted charge transfer is further confirmed by EIS analysis. The sulfur cathode with MTO-CNTs capping layer presents much lower charge-transfer resistance compared with cathode without MTO-CNTs capping layer, indicating that

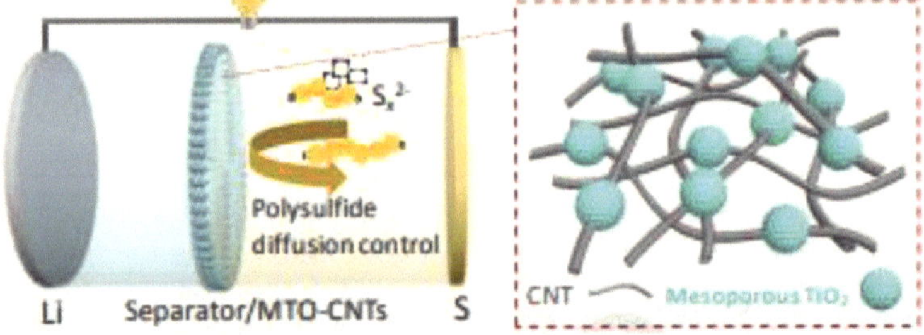

Figure 3.
Schematic diagram of as fabricated Li-S battery using MTO-CNTs interlayer. Reprinted from ref. [41].

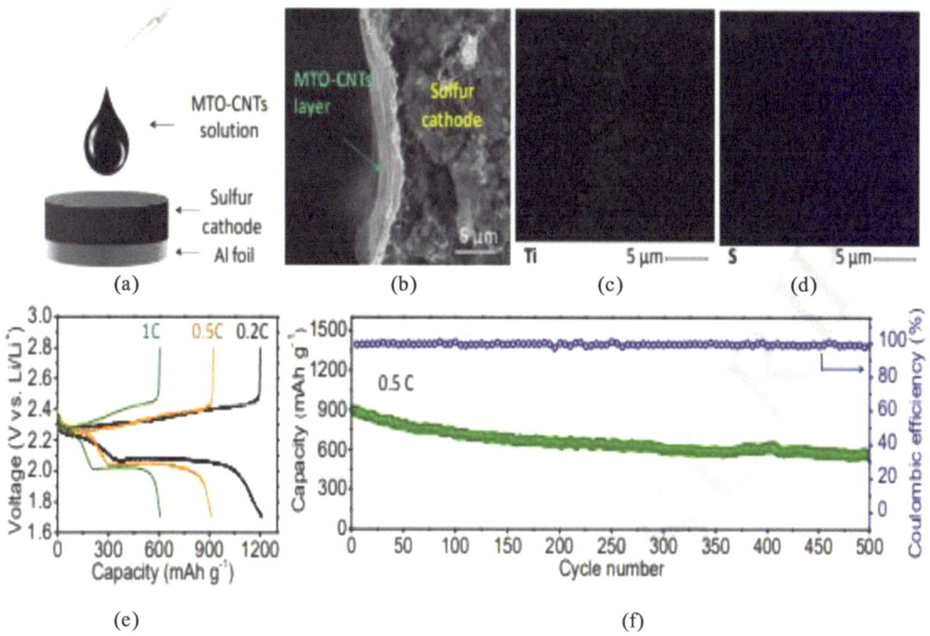

Figure 4.
(a) The dripping of MTO-CNTs on the surface of sulfur electrode. (b) SEM images of the sulfur electrode with MTO-CNTs capping layer. (c) Corresponding Ti and (d) S elemental mapping. (e) Charge-discharge curves of sulfur electrode comprises MTO-CNTs capping layer at various C rates. (f) Cycling stability of the MTO-CNTs capping layer on sulfur electrode at 0.5 C. [reprinted from ref. [41].

improvement of redox-conversion ability as well as conductivity. These results indicate that the capping layer is not only favorable for adsorption confinement of polysulfides within the electrode, but also increasing charge transfer, accelerating reversibility of polysulfide conversion. The galvanostatic charge-discharge profiles of the device with the capping layer were recorded at different rates as shown in **Figure 4e**. At the 0.2 C rate reversible capacity of 1212 mAh g^{-1} has been achieved. The specific capacities of 922 and 606 mAh g^{-1} were delivered at the high rates of 0.5 and 1 C, respectively. The cycling stability of the device with the capping layer at 0.5 C rate still it retains a capacity of 577 mAh g^{-1} after 500 cycles with capacity decay rate of 0.07% per cycle (**Figure 4f**), indicating a good cycling stability. These results indicate that the formation of MTO-CNTs capping layer is convenient route to fabricate high performance Li-S batteries with sulfur host along with commercial carbon materials.

The cell was disassembled after 100 cycles at 0.2 C, to understand the function of the MTO-CNTs capping layer at a potential of 2.8 V. The dimethyl carbonate solution was used to wash electrodes with capping layer and their structure analyzed by X-ray microtomography (XRM) and SEM. The overall structure and morphology of the MTO-CNTs are similar to that of the original sample as shown in **Figure 5a**, the SEM images of the capping layer after 100 cycles. The TiO_2 volume cannot change due to its robustness and the interaction of the CNTs. The layer-by-layer stacked structure was indicated by a 3D reconstruction of the electrode (**Figure 5b**). The capping layer is uniformly and strongly anchored on the sulfur cathode surface and worked as a good absorbent layer to keep polysulfide species rather than to diffuse into the lithium anode. The signal of sulfur precipitate is very strong and uniformed in the cathode (**Figure 5c**), reveals that the polysulfide shuttle behavior retained by the MTO-CNTs

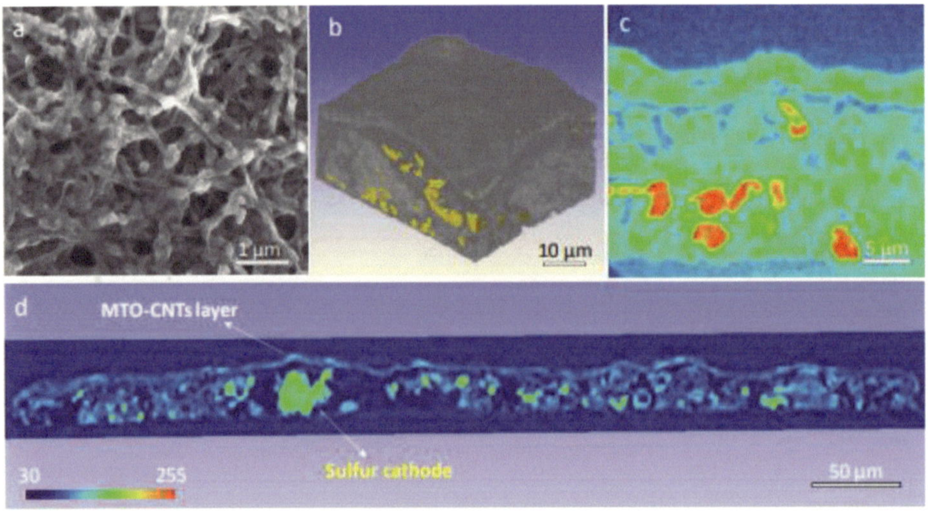

Figure 5.
(a) SEM picture and (b) 3D XRM picture and (c, d) partial 2D of the sulfur electrode with MTO-CNTs capping layer at a potential of 2.8 V after 100 cycles at 0.2 C. reprinted from ref: [41].

capping layer. The charge products are clearly seen in the capping layer execute the recycling of trapped polysulfides. **Figure 5d** reveals that the capping layer on surface of the sulfur cathode retain as it is without any cracks after cycling indicates that the structural stability. These results indicate that the light MTO-CNTs capping layer coated on the surface of sulfur cathode enhanced battery performances.

2.3 Multi (mixed cationic and anionic) redox in chalcogen cathode for Li-ion batteries

In order to improve the metal-ligand covalency by substituting a lower electronegative chalcogen ligand such as sulfur in the cathode. In this case, reversible mixed anionic and cationic redox occurs by the metal d band penetrates into the ligand p band. The partially filled d orbital redox couples like $Fe^{2+/3+}$ are introduced in the Li-ion conducting phase (Li_2SnS_3) is the development of a new family of layered structured cathode materials. Investigation of high-resolution transmission electron microscopy and high annular dark field-scanning transmission electron microscopy reveals that the multi-redox structural modifications and nanopore formation on its surface, during cycling process. In this study, Ni, Co ions free cathodes using various functional materials in the chalcogen-based dual anionic and cationic redox cathodes.

2.3.1 Electrochemical kinetic study

The charge-discharge profiles clearly exhibited a large hysteresis (0.511 V at 50% state of charge) starting from the first cycle. A voltage hysteresis plot for the 1st–30th cycles is shown in **Figure 6a**. Further to understand the hysteresis clearly by analyzed the effect of the upper cutoff potential on the voltage hysteresis as shown in **Figure 6b**. But these results have no significant effect on the voltage hysteresis, indicating that the hysteresis in the Sn-based chalcogen system may be due to the origin of intrinsic nature of the Li insertion extraction reactions. Therefore galvanostatic intermittent titration technique (GITT) measurements were carried out with current density of 5 mA/g for an interval of 30 min. Then it allowed opening circuit state for 1 hour, to obtain a steady state. Two continuous cycles of the composite electrodes by the GITT measurements are shown in **Figure 6c, d**. The first charge process represents the Low over potential increment gradually and the discharge process battery delivered a very high potential with different behavior. The 0.2 Fe substituted composite cathode shows the less overpotential with gradual increment is due to the smooth and kinetically faster of Li extraction process compared with the Li insertion process. The huge over potential observed around 2.5–2.1 V during the insertion process reveals that the Li insertion is kinetically limited and slower than the extraction process. In the similar way, the second cycle showed and high over potential during discharge and less over potential during charge condition. Further, the discharge profile region shows the two different slops indicating that due to the multiphase reaction region and the structural modifications. GITT profiles reveals that the inflection point region represents the high potential of ~187 mV compared to other regions. The electrochemical insertion of Li^+ ions is limited in these regions due to the slow Li-ion diffusion and charge transfer resistance.

The Li insertion and extraction was estimated by cycling test for all the compounds in a half-cell configuration at the current density of 10 mA/g. The electrochemical investigation of 0.2 Fe substituted compounds compared with other

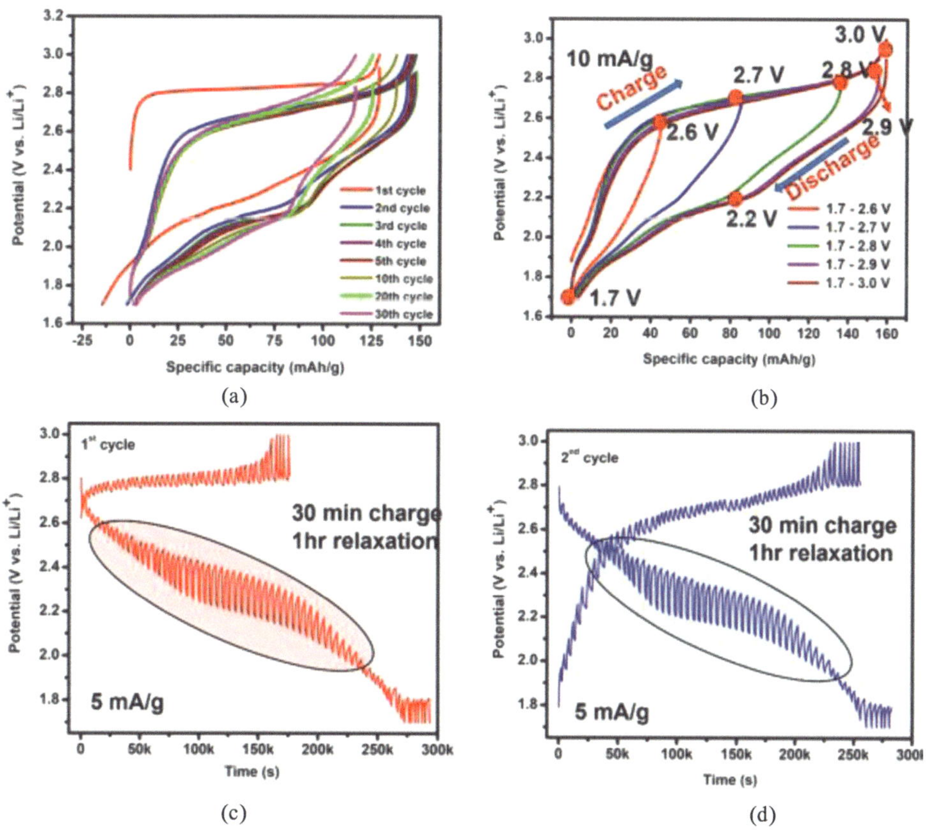

Figure 6.
Electrochemical kinetic study of the Sn-based chalcogen anion redox cathode ($0.2Fe\text{-}Li_{1.33}Sn_{0.67}S_2$). (a) Voltage hysteresis profile of active cathode material at 10 mA/g current density. (b) Voltage profile of the cathode at different upper cutoff potentials. GITT profiles of active cathode material: (c) 1st cycle and (d) 2nd cycle GITT profiles with 30 min pulse and 1 h relaxation. Reprinted from ref. [13].

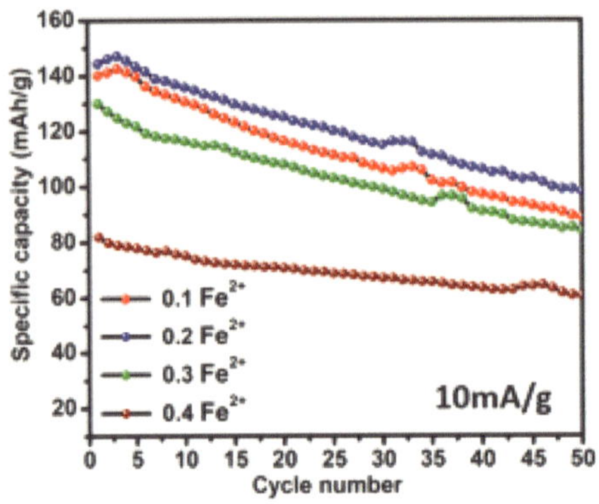

(a)

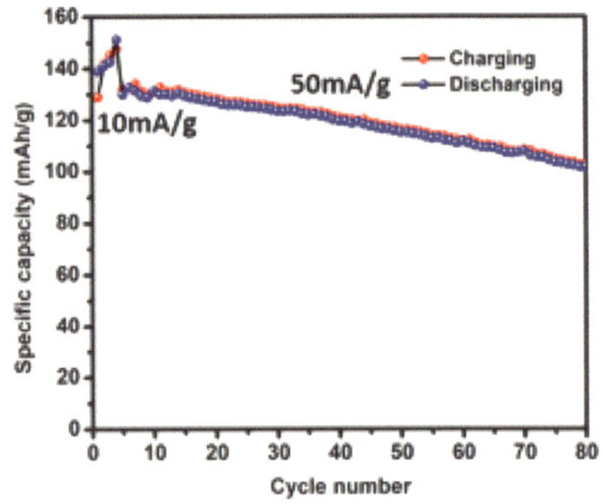

(b)

Figure 7.
Electrochemical study of Sn-based chalcogen anion redox cathodes. (a) Cycling stability of different Fe substituted $Li_{1.33}Sn_{0.67}S_2$compounds. (b) Cycling stability of 0.2Fe- $Li_{1.33}Sn_{0.67}S_2$ cathode at a high current density of 50 mA/g (initial few cycles at 10 mA/g) reprinted from ref. [13].

compositions are shown in **Figure 7a**. There is a gradual deterioration in all the compounds cyclic performance at a very low current density of 10 mA/g for 50 continuous cycles in terms of capacity fade. Further, a cycling test of the 0.2 Fe substituted cathode done at 50 mA/g, reveals that the high rate cycling stability about 76% retention after 80 cycles as shown in **Figure 7b**. The multi-redox induced structural transformation is the main reason for capacity degradation and evidenced by the microscopy analysis. The cycling stability of this composite materials is comparable to that of existing chalcogen anion redox cathodes [11, 56]. Further, the nanostructure and surface coatings strategies would enhance the cycle life [57, 58]. In this chalcogen framework the Fe doped composition showed good electronic and ionic conductivity, excellent electrochemical properties with the high loading of 10 mg/cm^2 cells at 10 mA/g current density.

2.3.2 HAADF-STEM and HR-TEM pictures

The structural evolution due to Li insertion extraction was visualized in a series of samples using HAADF-STEM images (**Figure 8**). High atomic number elements such as transition metals represented by the bright spots and the light elements such as Li, S, and O represented without bright spots. **Figure 8a** shows the ordered pristine cathode composition showed bright spots of metal elements and the lattice exhibits without distortions or cracks. The lattice showed severe structural distortions and nanopore formation after the 1st cycle, indicated by the yellow-colored circles and arrows in the **Figure 8c–f**.

The structural distortion was observed in the first discharged cycle like honeycomb and after consequent cycling the ordered crystalline domain lost its crystalline nature as shown in **Figure 8c**. Further, investigations reveal that complete distorted structure after cycling due to the pores and structural distortions was high and the crystallinity degraded. The degradation of pore and crack formation happened during the initial cycles to extended cycling conditions. The low magnification images of second charged, second discharged, and cycled samples are shown in **Figure 8g, h** indicating a lot of nanopores present throughout the cycled cathode. Hence, the nanopore created by accelerated Li ion insertion-extraction

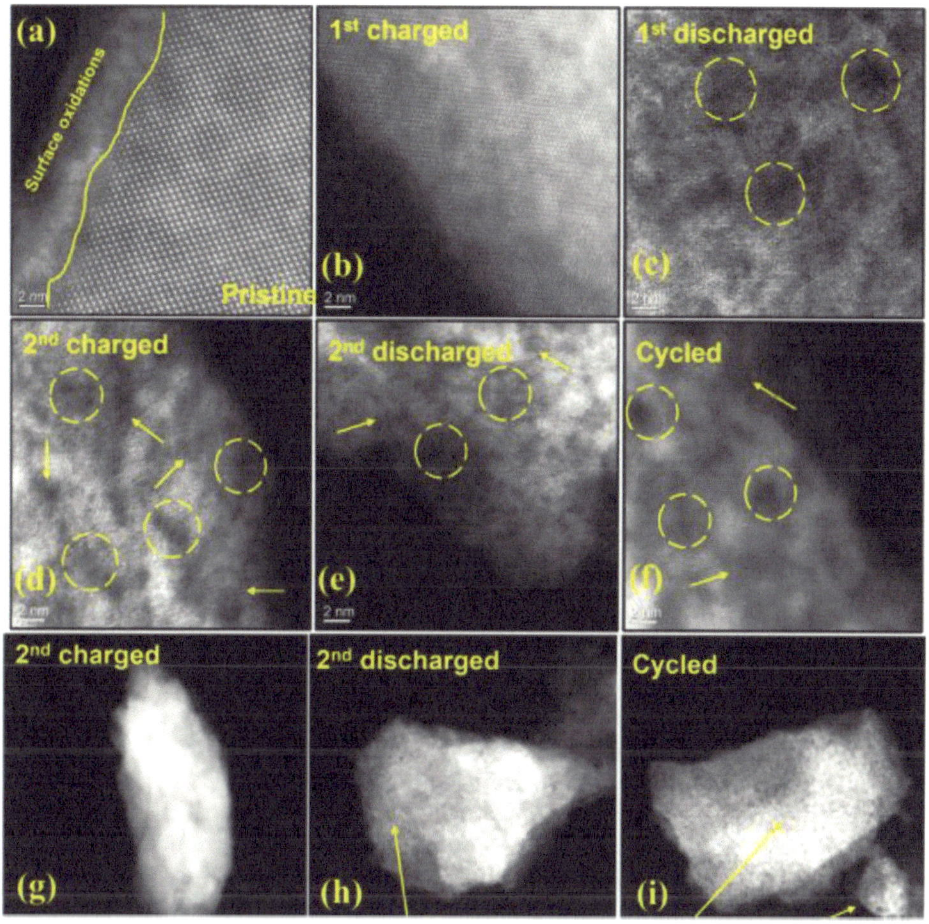

Figure 8.
HAADF-STEM investigation of the 0.2Fe-$Li_{1.33}$Sn0.67S_2composite electrode at different states of charge: (a) pristine, (b) 1st charged, (c) 1st discharged, (d, g) 2nd charged, (e, h) 2nd discharged, (f, i) cycled electrodes; In panel (c), the honeycomb ordering was visualized. Reprinted from ref. [13].

and also the sulfur loss by degradation. The formation of nanopore/nanovoid due to oxygen loss in the Li-rich oxide anion redox cathode as well as their degradation mechanisms correlated with the fundamental issues of voltage fade, voltage hysteresis, and capacity fade [59]. The sulfide anion is a significant charge contributor during Li extraction and the sulfur loss increased amorphization which is reflected in charge-discharge voltage profiles by means of capacity fade of the cycled cathode materials.

2.3.3 Impedance spectroscopy analysis

Figure 9a shows the impedance plots of cathode material at different discharged cycles. All the EIS measurements showed two semicircle regions except the pristine cell. The semicircle observed at high frequency region indicates the surface film formation on the positive electrodes. Another semicircle was observed in the low frequency region attributed to the charge transfer resistance upon Li^+ insertion-extraction. The observed slope line indicates the Warburg diffusion (W) in the bulk electrolyte. An equivalent

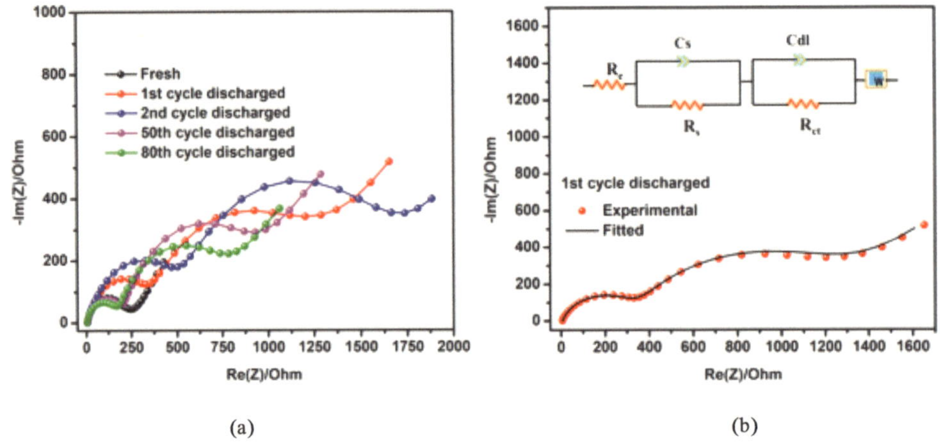

Figure 9.
EIS of 0.2Fe-$Li_{1.33}$ $Sn_{0.67}S_2$ composite electrode at different states of charge. (a) Nyquist plots of the composite electrode at various cycles. This EIS was recorded after completion of respective cycle, (b the model EIS was fitted to the experimental Nyquist plot. Reprinted from ref. [13].

circuit model (**Figure 9b**) designed by the fitted EIS data and the resistance values for all the active materials at different cycles noted. In the equivalent ckt R_e corresponds to the solution resistance. The Cs and Rs correspond to and capacitance and the surface film resistance of the active sulfide composite cathode, respectively. The Li ion insertion and extraction double layer capacitance and induced charge transfer resistance represents as R_{ct} and C_{dl}, respectively. From the **Figure 9a** calculated R_e values for all the materials indicated small differences for all the cycles, reveals that the electrolyte is stable during cyclic process. The R_s and R_{ct} values increased proportionally in the initial cycle number during 1st and 2nd cycles, representing that the structural distortion during the electrolyte and electrode interface. This observation consisted with GITT analysis that the very large potential during lithium insertion of initial cycles is due to the kinetic limitation of Li insertion reactions. The most of the surface film formation will occur during initial cycling. The R_{ct} values decreased with increasing cyclic number, indicating that the resistance increased during the initial cycles and decreased by cycling cyclic number. This is due to the sulfur loss by rising cycling number, where the complete amorphization and nanopore are formed [12]. The anion redox indicates sulfur loss by reducing the Li insertion-extraction ability. These indicate the capacity fading occurred during the cycling. Therefore Sn based chalcogen layered structure materials worked as mixed redox cathode for Lithium Ion batteries.

2.4 Chalcogen substitusion into the polysulfides for batteries

The cyclic voltammograms (CVs) of half cells assembled using [Li_2S + 0.1 S/Se/Te] cathodes as shown in **Figure 10a**. The presence of polyselenosulfides plays a significant reduction in peak separation (ΔE_p), indicating decreasing the overpotentials, and helps with increased peak heights at high scan rates ($1 \geq$ mV s^{-1}) by retaining the canonical redox peaks of sulfur/Li_2S. The relationship between the peak current (i_p) and scan rate (ν) can be written as: $i_p = \alpha \nu^{\beta}$, where α and β are the fitting parameters [60] Plotting log (i_p) versus log (ν) yields β = 0.64 for Se, compared to 0.52 for the control. The value of β changes with the addition of Se represents the shifting of slow diffusion controlled reactions to fast surface-controlled reactions. This improvement of the redox kinetics is not much with the introduction of tellurium compared with selenium. **Figure 10b** shows the capacities of Li || [Li_2S + 0.1 S/Se/Te] half cells at 0.25 A g^{-1} of Li_2S (BC/5). Selenium is electrochemically more active between 2.8 and 1.8 V, whereas Te is inactive at the same voltage. Hence the addition of 0.1 Se enables a considerable improvement in capacities of ~40% under the control. Therefore the relative dominance of catalytic SeS_2^* radical presence in polyselenosulfide solutions works as conversion reactions and utilized completely to drives complete electrochemical reaction [61, 62]. This is critical for high capacities obtained under low-electrolyte conditions in a practical Li–S cell [24, 63]. The presence of SeS_2^* radicals highly react with the various electrolyte components thereby the faster capacity fade observed with polyselenosulfides even after 70 cycles. The sulfur/Li_2S final product conductivity also improved due to incorporation of Se atoms. The charging/discharging profiles of Li || Li_2S half cells indicate that considerable reduction achieved in overpotentials with selenium compared to sulfur or tellurium. Therefore, considerable improvement in charge-transfer and redox kinetics is occurred with the introduction of selenium rather than tellurium.

2.4.1 Impact of Se and Te substituted polysulfides on lithium deposition

Electrochemical performance of Li_2S and Li_2S_2 entirely depends on the lithium plating and stripping effectively. **Figure 11a** shows capacities for anode-free Ni || [Li_2S + 0.1 S/Se/Te] full cells at ~ 1 mA cm^{-2} (C/5). Selenium showed rapid capacity fade and loss of their peak capacity 50% after the 35 cycles. By the introduction of tellurium exhibited remarkable cycling stability in the anode-free configuration

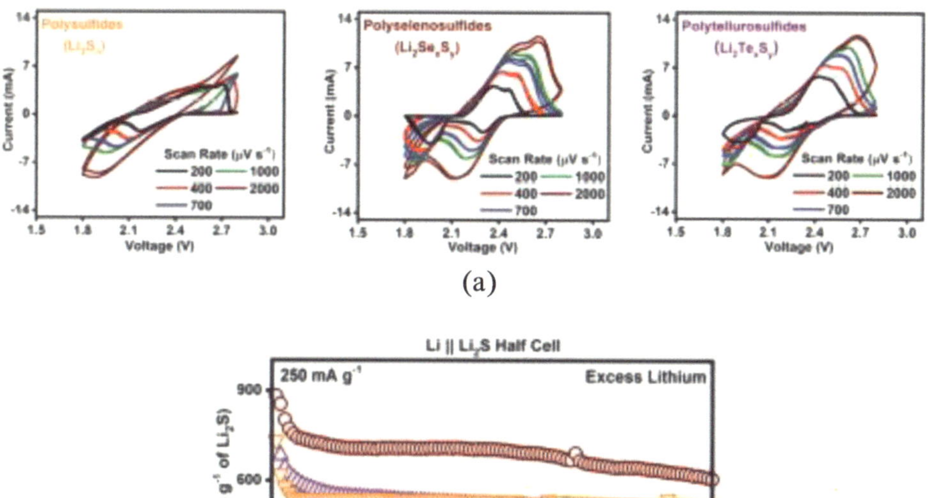

(a)

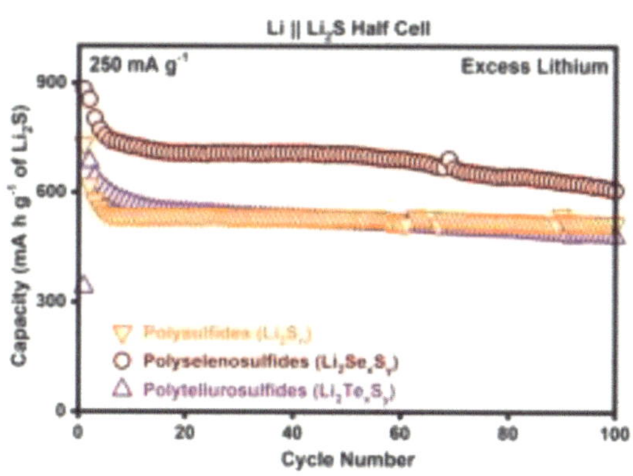

(b)

Figure 10.
(a) Cyclic voltammograms for [Li_2S + 0.1 S/Se/Te] cathodes at scan rate range of 200 to 2000 μv s^{-1} (b) electrochemical performance of Li || [Li_2S + 0.1 S/Se/Te] half cells. Reprinted from ref. [26].

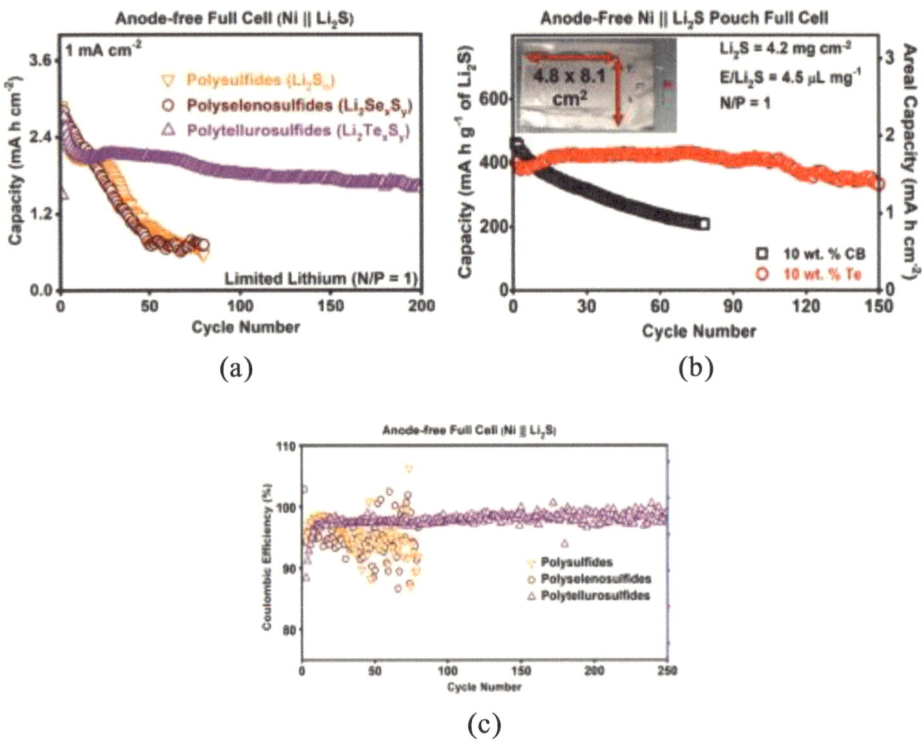

Figure 11.
Electrochemical performances – (a) capacity retention and (b) coulombic efficiencies of anode-free Ni || [Li_2S + 0.1 S/Se/Te] full cells. (c) Electrochemical performances of large-area (39 cm^2) anode-free Ni || Li_2S single-layer pouch full cells with 10 wt% tellurium (Te: Li_2S molar ratio = 0.04) or 10 wt% carbon black as cathode additives. Reprinted from ref [26].

and retain 52% of peak capacity at 265 cycles. The loss rate of lithium per cycle is decreased to 2.14% with Se and 0.24% with Te [64]. The improvement in lithium plating and stripping reversibility reflects by the coulombic efficiencies of the anode-free full cells as shown in **Figure 11b**. Polytellurosulfides showed a dramatic effect on lithium cycling efficiency by situ formation, kinetic hindrance occurred with tellurium substitution in polysulfides compared with selenium. The formation of polyselenosulfides has no effect on the reversibility of lithium deposition. These improvements were analyzed with symmetric Li || Li cells containing $Li_2Se_xS_y$ and $Li_2Te_xS_y$ introduced as an electrolyte components. Electrochemical impedance spectroscopy reveals the polyselenosulfides showed high and unstable overpotentials (~100 mV) and polytellurosulfides enable low and stable overpotentials (~10 mV), indicating a thin SEI layer has excellent ionic transport properties. The dense and uniform lithium deposits formed with polytellurosulfides exhibited irreversible loss of lithium. The electrochemical performance of an anode-free Ni || Li_2S full cell with 0.05 Se + 0.05 Te additives was observed. There is a synergetic effect realized that the higher initial capacities and cycling stability than that of the pure 0.1 Te and 0.1 Se based cells. Therefore we believe that the presence of SeS_2*radicals increased the faster capacity fade with 0.05 Se + 0.05 Te than with 0.1 Te, representing that an electrolyte system might be allowed radical anion of selenium to obtained higher capacities and also retained for longer cycles. These results further implemented impractical, large-area (4.8 × 8.1 cm^2), single-layer pouch cells assembled in the

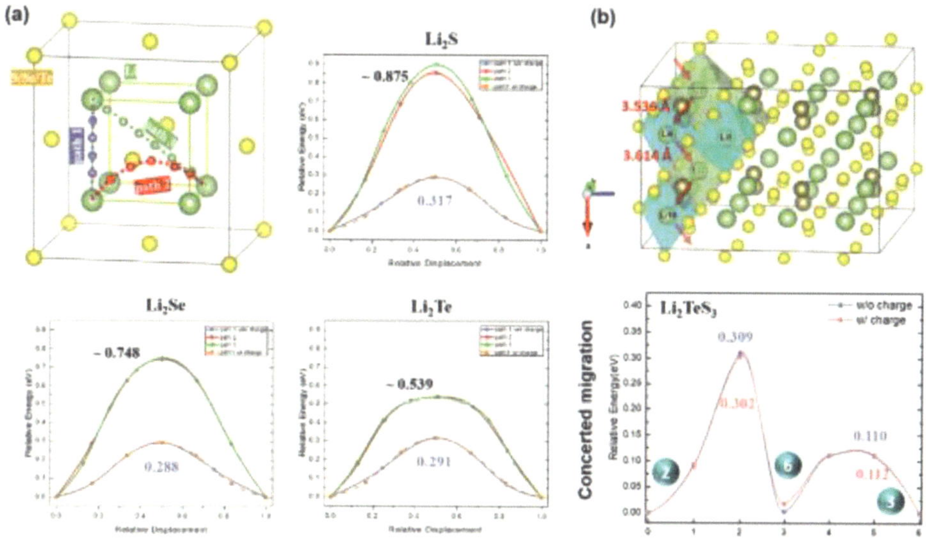

Figure 12.
(a) Crystal structure of Li_2X (X = S, Se, and Te) and the three Li^+ ion diffusion pathways marked as purple [100], red [110], and green [111] lines. Migration energy barriers along [110, 111] show a steady reduction from Li_2S to Li_2Se and to Li_2Te. (b) Li^+-ion transport pathway in Li_2TeS_3 along the x-axis and the corresponding energy barrier based on single-ion migration. Reprinted from ref [26].

anode-free configuration (N/P = 1) with a 164 mg Li_2S cathode (4.2 mg cm^{-2}) containing 10 wt% Te^0 (Te:Li_2S molar ratio = 0.055) and operating under lean-electrolyte conditions (E/Li_2S = 4.5 ml mg^{-1}) the results are shown in **Figure 12c** [65]. Tellurium was replaced with carbon black for control. **Figure 11b** showed the control cell exhibited a high initial capacity of 77 mAh, but it has very rapid capacity fade 80% retention within 13 cycles. By the addition of Te exceeded 80% of its peak capacity for nearly 150 cycles and retain their cycling capacity without rapid drop until the electrolyte dry-out nearly 300 cycles [66, 67]. The initial rise in capacity can be regarded as 'activation period' in which the dissolution of tellurium slowly into polysulfides. The improvement in cycle life with the introduction of tellurium can be attributed to the stabilizing effect of polytellurosulfides on lithium deposition. These results are valid practically relevant to the cell design and testing parameters such as long cycle life and high energy dense, anode-free configuration significantly closer to commercial viability of Li–S system.

2.4.2 Lithium-ion transport properties of selenides, tellurides, and thiotellurates

First principles calculations were evaluated to understand reduced species on lithium deposition and their ionic transport properties. Li_2S, Li_2Se, and Li_2Te indicated crystallize in a cubic antifluorite structure with a face-centered cubic anionic framework along with Li^+ ions in the tetrahedral sites. **Figure 12a** showed Li^+ can diffuse along the directions of [100], [110], [111]. Climbing image nudge elastic-band (CI-NEB) method is used to find the diffusion barriers along each of these pathways. Barrier energy found to be ~0.3 eV in the lowest- pathway [68]. Barrier energies are 0.875 eV to 0.748 eV to 0.539 eV, calculated from the transitions Li_2S to Li_2Se to Li_2Te, respectively. Te_2 exhibited more polarizable anionic framework compared to those of $S_2$and Se_2due to the larger size and lower charge density. Previous report [69] reveals that the larger size of Te_2 provide more open channel along [110], [111] in which high diffusion pathways for Li^+ ions. Therefore, Li_2Te due to its more uniform, homogenous, and dense lithium deposition can provide alternate pathways to facilitate three-dimensional ion transport. Li_2TeS_3 exhibited monoclinic structure with trigonal pyramidal TeS_3^{2-}anions arranged in layers. The Li^+ ions coordinated with sulfur atoms and occupied their octahedral and tetrahedral sites. Li_2TeS_3 unit cell consists of eight distinct steps between five adjacent sides and the non-equivalent lithium sites can be found in the migration pathway. The single-ion NEB model was introduced to calculate the corresponding barrier energies and find the most favorable path. It indicates the migration from one tetrahedral site to another tetrahedral site through an intermediate octahedral site in the direction of the x-axis as shown in **Figure 12** (b). The migration barriers are found to be 0.378 and 0.250 eV. The barrier energies are found to be 0.4 and 0.6 eV for other migration pathways. Therefore, multiple viable Li^+ ion diffusion pathways available in Li_2TeS_3 in three dimensional paths ways for ion transport due to stable and reversible lithium deposition. All these factors would improve the lithium cyclic efficiencies by the formation of interfacial components with polytellurosulfides.

2.4.3 Interfacial chemistry of Se and Te substituted polysulfides on lithium batteries

In order to understand their effects of modified Li_2Sn, Li_2Se_xSy and Li_2Te_xSy species by XPS, in which lithium anode-free full cells analyzed after 20 cycles. **Figure 13a** shows the S 2p + Se 3p and Li 1 s + Se 3d spectra for the cell with addition of 0.1 Se. The S 2p spectra $SO_4{}^{2-}$ species are dominated by oxidized sulfur due to the decomposition of LiTFSI. The dominated peaks at 165 eV and 58.7 eV are appeared in the Se 3p and Se 3d spectra, respectively by the oxidized Se^{+4} in selenites (SeO_3^{2-}) and minor components are present due to Reduced sulfur species (Li_2S). The presence of oxidized selenium species is due to $LiNO_3$, which is a strong oxidizing agent and oxidizes selenides (Se_2) into selenites ($SeO_3{}^{2-}$) [70]. Therefore, polyselenosulfides introduced not only the fundamental alter to the lithium–electrolyte interface, which remains dominated by oxidized sulfur/selenium species. **Figure 13b** shows the cell with 0.1 Te additive spectra of S 2p and Te 3d. The reduced sulfur species (S^{2-} at 160.6 eV) exhibited dominated S 2p spectra. Likewise, the Te 3d spectra are dominated by sulfurized tellurium species (Te^{+4} at 574.6 eV). The quantification of the spectra reveals the formation of thiotellurate (TeS_3^{2-}) species [70]. Thus, the formation of Li_2TeS_3 as the dominant interfacial component and are reduced on the lithium surface. Some of previous research reveals that the oxidized sulfur species are present as minor components. It is extended that the oxidized tellurium species ($TeO_3{}^{2-}$) made only a minor fraction of tellurium atoms on the lithium surface. Hence introduction of tellurium alters the lithium–electrolyte interface by the reduction of sulfur species (as Li_2TeS_3). These XPS observations consistent with the time-of-flight secondary ion mass spectrometry (ToF-SIMS). **Figure 13c** shows the profiles for Li_2 (metallic lithium) and SO_3 for three-dimensional reconstructions. A thick layer of electrolyte decomposition products is observed on the deposited lithium with polyselenosulfides but not with polytellurosulfides. Depth profiles indicate that the selenium has strong signal for SeO is compared to that for SeS. This is reversed for tellurium, in which TeS exhibited much stronger compared to that for TeO. Thus, majority of tellurium atoms made bond with sulfur and the majority selenium atoms making bond with oxygen. This is due to oxidation of $LiNO_3$ explained by Pearson's HSAB theory [71]. The soft Lewis acid cations (Te^{+4}) are formed by

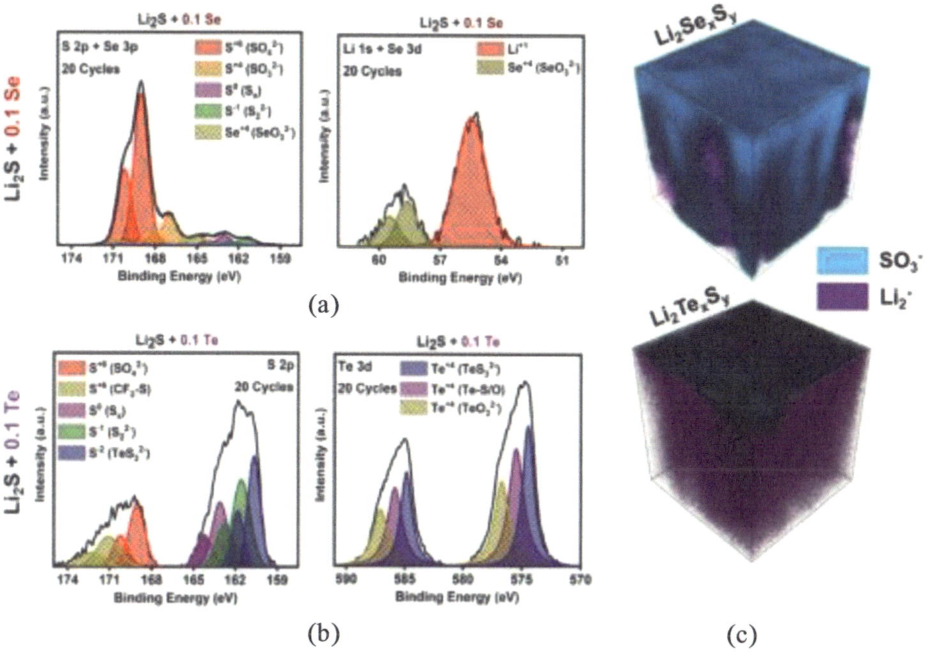

Figure 13.
(a) S 2p + Se 3p and Li 1 s + Se 3d spectra for the lithium surface in an anode-free full cell cycled with polyselenosulfides. (b) S 2p and Te 3d spectra for the lithium surface in an anode-free full cell cycled with polytellurosulfides. (c) 3D reconstructions of ToF-SIMS depth profiles for Li_2 (metallic lithium) and SO_2 (oxidized sulfur species) secondary ions. Reprinted from ref. [26].

the Tellurium that prefer soft Lewis bases such as S_2 sulfides, while selenium forms hard Lewis acid cations (Se^{+4}) that prefer hard Lewis bases such as O_2oxides [72, 73]. The divergent lithium stabilization capabilities of polyselenosulfides and polytellurosulfides explained the differences in lithium interfacial chemistry. The sulfide anionic framework such as Li_2TeS_3 identified as preferable compared to an oxide anionic framework such as Li_2SeO_3 or Li_2SO_3 [74]. The greater size and polarizability of S_2compared to those of O_2, improves ionic transport properties by reducing Li^+ ion diffusion barriers. The varying compositions of tellurium and selenium to get a stable sulfide-rich SEI layer, in the presence of $LiNO_3$ changed the characteristics of lithium deposition [75].

2.5 Freestanding $Se_{1-x}S_x$ foamy cathodes for high-performance Li-$Se_{1-x}S_x$ batteries

The development of supercritical CO_2 synthesis of selenium-sulfur solid solutions ($Se_{1-x}S_x$) are promising new cathodic materials for high-performance secondary lithium batteries due to their high electric conductivity than S and superior theoretical specific capacity than Se. The morphology and microstructure of N-doped carbon framework with three-dimensional (3D) interconnected porous structure (NC@SWCNTs) host are characterized by SEM and TEM pictures as shown in **Figure 14**. A depicted in **Figure 14a** shows the NC@SWCNTs host 3D honeycombed structure and interconnected melamine foam framework. The magnification SEM images (**Figure 14a, c**) reveals that numerous interlaced SWCNTs are covered the surface of melamine foam by the derived carbon skeletons and SWCNTs are formed as thin sheets between carbon skeletons. This structure of NC@ SWCNTs exhibited a highly conductive 3D network to transport the electron or ion, but also increases the mechanical strength as well as flexibility of NC@SWCNTs host. TEM results (**Figure 14d**) reveals that SWCNTs, 3D network structure are crisscrossed in carbon skeletons. By EDS analysis the main elements found to be in the NC@SWCNTs are C, O and N, which are uniformly distributed as shown in **Figure 14e**. N signal is derived from melamine foam because of it contain high N. 3D network structure of NC@ SWCNTs is made of by composing SWCNTs-coated N-doped carbon skeleton melamine foam and wafery sheets of SWCNTs. The NC@SWCNTs consists of pores and layer gaps are favorable for loading of $Se_{1-x}S_x$ active conductive materials. The 3D conductive network promotes not only redox kinetics, but also endow NC@ SWCNTs host with strong buffering in volume during cycling. Further, N doped is also beneficial for the adsorption of intermediates, after $Se_{1-x}S_x$ impregnation, compared to NC@SWCNTs host. NC@ SWCNTs@ $Se_{1-x}S_x$ composites retain their original morphology of NC@SWCNTs and no

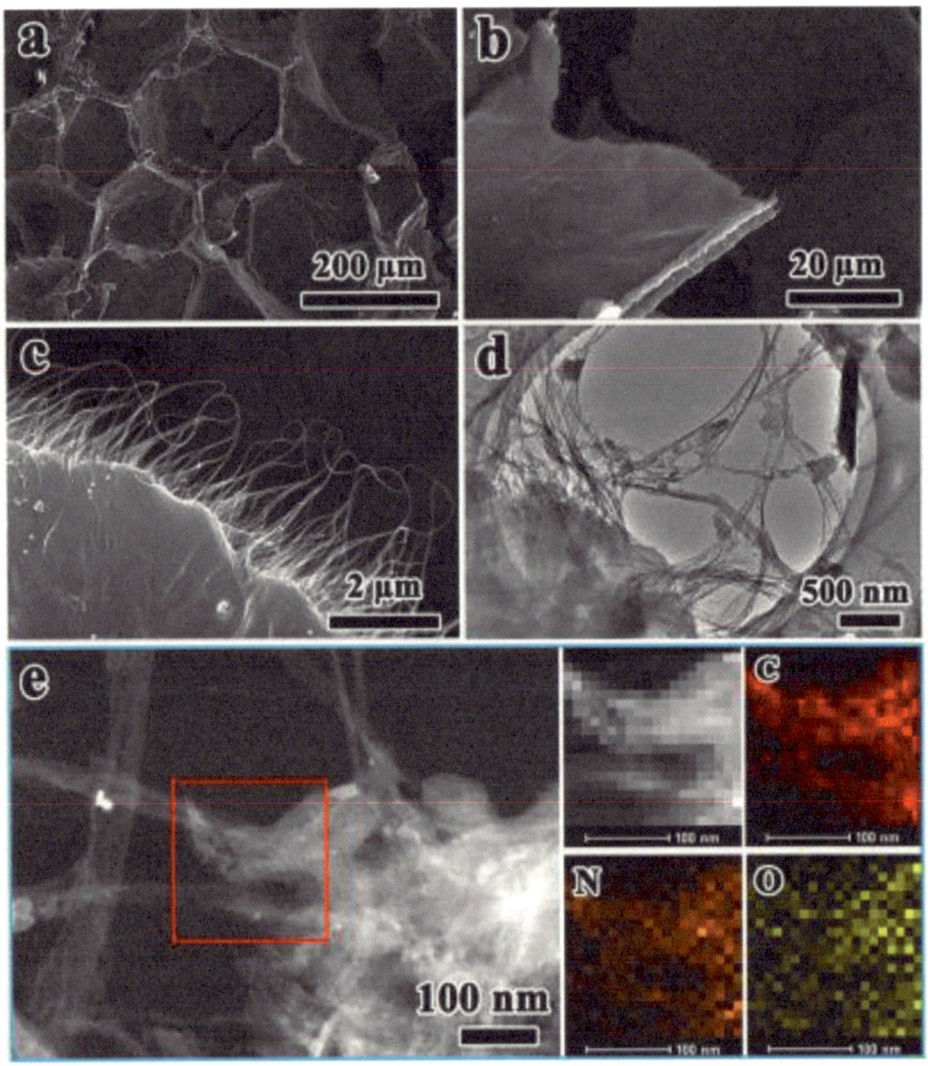

Figure 14.
(a–c) Shows the SEM images, (d) represents the TEM image (e) indicated STEM image of NC@SWCNTs and the corresponding mapping images. Reprinted from ref. [76].

discernible $Se_{1-x}S_x$ particles can be found. According to EDS results, the C, N, Se and Se signals are overlapped well, suggesting $Se_{1-x}S_x$ composites are uniformly permeated into the pores and layer gaps of NC@SWCNTs host with the assistance of SC-CO_2 due to the good permeability, excellent diffusivity and high solubility of SC-CO_2 [76].

Electrochemical performance of NC@ SWCNTs@ $Se_{1-x}S_x$ composites based cathodes evaluated in Li- $Se_{1-x}S_x$ batteries using carbonate-based electrolyte ($LiPF_6$-EC/DMC). **Figure 15a** show initial three cyclic voltammetry (CV) curves of NC@SWCNTs@ $Se_{1-x}S_x$ cathodes with scanning rate of 0.1 mV s^{-1} in the potential range from 1.0 to 3.0 V versus Li/Li^+. Initially, a sharp reduction peak at ~1.38 V, a small reduction peak at ~2.37 V and a broadened oxidation peak at ~2.14 V appeared. The sharp reduction peak at ~1.38 V shifts to ~1.7 V and the small reduction peak at ~2.37 V was disappeared after the first scan. Initially, during the lithiation the peak shift due the activation process and the polarization is also further reduced [77] CV curves overlapped after the first scan reveals that the good cyclability and reversibility of NC@SWCNTs@ $Se_{0.2}S_{0.8}$ cathode [78]. Notably, CV curves of NC@SWCNTs@ $Se_{1-x}S_x$ cathodes obtained differently from S cathode, representing the change of electrochemical reaction of S by Se and it is more conducive and stable with carbonate-based electrolytes. **Figure 15b** shows the galvanostatic charge-discharge curves of NC@SWCNTs@ $Se_{1-x}S_x$ cathodes are consistent with the result of CV. There are two plateaus are observed (1) extremely short plateau at ~2.38 V, and (2) a long plateau at ~1.75 V in the first cycle. Further, subsequent cycles, long plateau at ~1.75 V becomes a little steeper and shifts to ~1.88 V and the short plateau appeared at ~2.38 V. The short plateau appeared

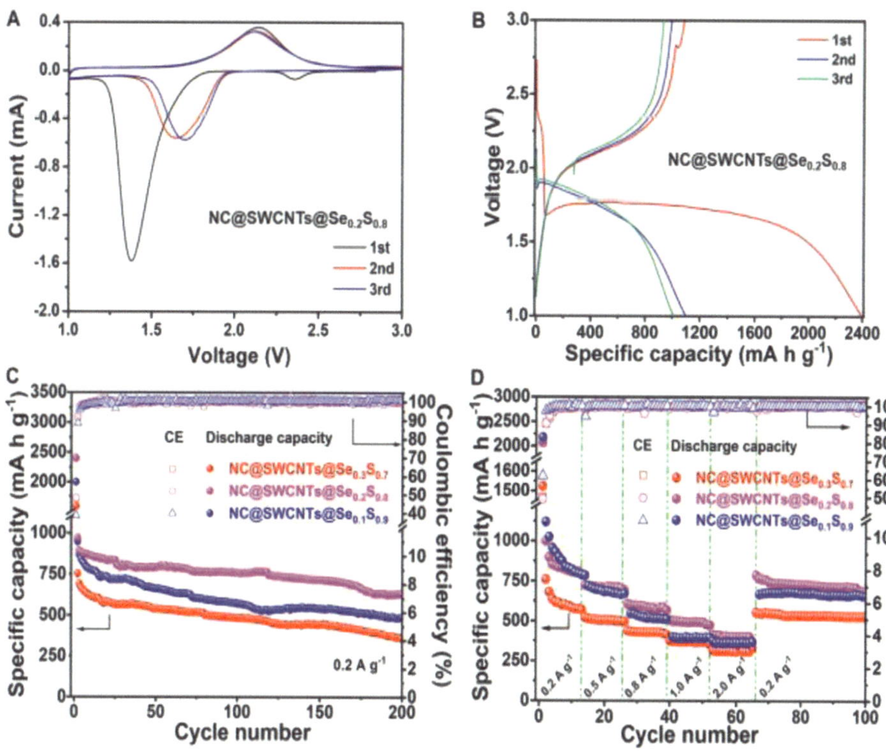

Figure 15.
(A) CV curves of NC@SWCNTs@$Se_{0.2}S_{0.8}$ cathode. (B) Charge/discharge curves of the NC@SWCNTs@ $Se_{0.2}S_{0.8}$ cathode at 0.2 a g^{-1}. (C) Cycle performances and (D) rate performances of NC@SWCNTs@$Se_{1-x}S_x$ cathodes. Reprinted from ref [76].

at ~2.38 V is attributed to the transformation of $Se_{0.2}S_{0.8}$ into polysulfides/polyselenides. The short plateau is disappeared due to dissolution of intermediates into the electrolyte [79]. The long plateau at 1.75–1.88 V is attributed to the conversion of polysulfides/polyselenides to Li_2S/Li_2Se [77]. There is only one sloping plateau appeared during the charge process, at ~2.12 V due to the conversion of Li_2Se/Li_2S to $Se_{0.2}S_{0.8}$. The cyclic performance of NC@SWCNTs@ $Se_{1-x}S_x$ cathodes at a current density of 0.2 A g^{-1} with different Se/S ratios as shown in **Figure 15c**. As prepared NC@SWCNTs@$Se_{0.2}S_{0.8}$ cathode delivers the highest initial discharge capacity (2398.5 mA h g^{-1}) among all the samples. Discharge capacity exceeds the theoretical capacity at initial stage may be due to side reactions and the formation of SEI layer on the surface of electrode [80]. Electrochemical characteristics of NC@SWCNTs@ $Se_{0.2}S_{0.8}$ cathode exhibit the superior cyclic stability. **Figure 15d** showed the rate capabilities of NC@SWCNTs@ $Se_{1-x}S_x$ cathodes at various current densities. Among all the samples, NC@SWCNTs@ $Se_{0.2}S_{0.8}$ cathode showed the best rate performance. At the various current densities of 0.2, 0.5, 0.8, 1.0 and 2.0 A g^{-1} the reversible rate capacities of NC@SWCNTs@ $Se_{0.2}S_{0.8}$ cathode are found to be 998.4, 723.7, 606.8, 506.1, and 415.0 mA h g^{-1}, respectively. The reversible discharge capacity of NC@SWCNTs@ $Se_{0.2}S_{0.8}$ cathode reverts to the initial value, when the current density switches back to 0.5 A g^{-1}. NC@SWCNTs@ $Se_{0.2}S_{0.8}$ cathode with Se loading of as high as 4.4 mg cm^{-2} exhibited areal capacity of as high as 2.78 mA h cm^{-2} is the best candidate most reported $Se_{1-x}S_x$ cathodes in literature [77, 81–83]. The electrochemical performance of NC@SWCNTs@ $Se_{0.2}S_{0.8}$ cathode is more effective due to the following reasons: 1) Se and S in $Se_{0.2}S_{0.8}$ solid solution play various roles: Se can improve more electrical conductivity, whereas the S can raise its capacity. 2) N-doped 3D porous carbon matrix and interlaced SWCNTs can provide storage and the structural stability; thereby promote the cycling stability of NC@ SWCNTs@ $Se_{1-x}S_x$ cathodes. NC@SWCNTs@ $Se_{0.2}S_{0.8}$ cathode exhibits good cycling stability (632 mA h g^{-1} at 0.2 A g^{-1} at 200 cycle) and high rate performance (415 mA h g^{-1} at 2 A g^{-1}) due to well-designed structure as well as optimized chemical composition with in carbonate-based electrolyte. Hence these developments of high-performance $Se_{1-x}S_x$ cathodes suitable for advanced Li- $Se_{1-x}S_x$ batteries.

2.6 Nanoreactors for metal-chalcogen batteries

Porous hollow nanoreactors are investigated widely for lithium selenium and tellurium batteries. The mesoporous material exhibited considerable porosity (0.2 cm^3 g^{-1}) and a large surface area of 462 m^2 g^{-1}, which allowed for uniform distribution of Se8. The Se8/C based lithium selenium batteries showed a high reversible capacity of 480 mA h g^{-1} at 0.25C (1C = 678 mA g^{-1}) without loss of its capacity after 1000 cycles [84]. Further development of the Se/porous carbon cathode battery showed a high volumetric capacity of 3150 mA h cm^{-3} with excellent rate capability about 1850 mA h cm^{-3} at 20C. Therefore, it will be used for future commercialization of LSeBs [85]. Single-atom Co decorated hollow porous carbon also works as a nanoreactor with superior catalytic activity to polyselenides. These (Se@CoSA-HC) cathodes based batteries exhibited high discharge capacity, superior cycling stability,an excellent rate capability [86]. Metal or heteroatom doping (N, S, and Co) is also another alternative approach to enhance the utilization of Se or Te [87–89]. He et al. synthesized a nanoporous Co and N-co-doped carbon nanoreactor (C–Co–N) provide a high Te loading (77.2 wt%) provide ultrahigh capacity of 2615.2 mA h cm^{-3} and superior rate performance of 894.8 mA h cm^{-3} at 20C as shown in **Figure 16** [90]. Design structure and micro-environmental of Te-based nanoreactors could provide high

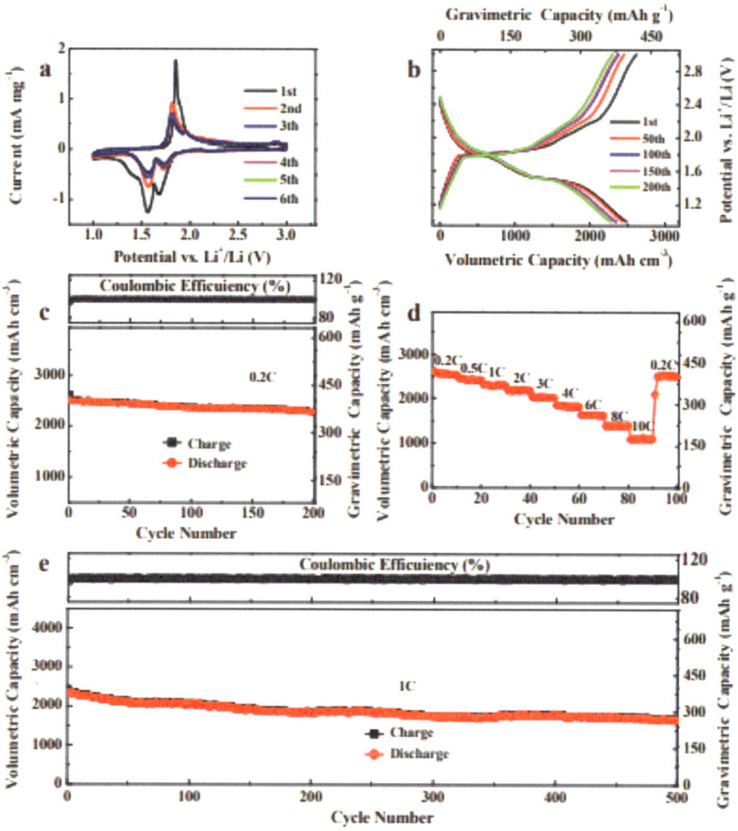

Figure 16.
(a) CV curves of the three dimensional rGo/tellurium (3DGT) aerogel at a scan rate of 0.1 mV s⁻¹. (b) Discharge curves of the 3DGT cathode at a 0.2 C. (c) Cyclic performance of the 3DGT cathode at 0.2 C for 200 cycles. (d) Rate performance at various crates for the 3DGT cathode. (e) Cyclic stability of 3DGT cathode at 1 C for 500 cycles. Reprinted from ref. [90].

electrochemical performance. In conclusion, the development of hollow porous nanoreactors not only provide a suitable specified space for chalcogens (S, Se, and Te), but also load active species for the regulation of the microenvironment in the electrode. Further development of nanoreactors, it is necessary to design the new methodologies at the molecular level to regulate the microenvironment of the catalyst.

2.7 Flexible cholcogen lithium-ion batteries

2.7.1 Flexible sulfur cathodes

Zhang and his coworkers developed 1D/3D hybrid flexible sulfur electrodes with good flexibility and exhibited improved electrochemical performance [91, 92]. They used sulfur-infiltrated 3D nanostructure porous carbon materials with various sizes nanometers to ten micrometers representing with high versatility and applicability for constructing flexible electrodes. These materials not sustain without support, therefore by incorporating ultralong CNT scaffolds, very robust films are obtained without sacrificing mechanical flexibility compared to ultralong CNT/MWCNT film. Such a materials exhibited tremendous specific surface area, high micro or mesoporosity, and surface functionalities than MWCNTs. Hence, this strategy is an ideal generic and versatile host to facilitate flexible sulfur cathodes. The use of graphene in 2D/3D hybridized structure is essential to alternative of CNTs in the 1D/3D, which provides required mechanical adhesion and good electrical conduction into 3D carbon constituents, but that lack of flexibility. Therefore, Wu et al. [93] demonstrated freestanding graphene based hierarchical porous carbon (GPC) films for flexible sulfur cathodes batteries as shown in **Figure 17**. Graphene-based microporous carbon (GMC) sheets are obtained by thin layers of microporous carbon were coated on both sides of GO after hydrothermal carbonization and KOH activation. The small sulfur molecules are stored in rich micropores of GMC, provides stronger physical confinement than normal graphene. Therefore, GPC files based batteries showed excellent cycling performance with stabilized capacities of 1030(422) and 626(357) mA h $g_{sul}(ele)^{-1}$ at 0.2C with the sulfur content as 41 and 57 wt%, respectively. In general, graphene-based film electrodes showed rapid decay in their capacity due to their polysulfide dissolution. Furthermore, the GPC-S cathode films used in flexible Li-S batteries by attaching the tape to pack the material, displaying comparable electrochemical performance in both flat and bent states as shown in **Figure 18**(a) and (b). Ni et al. [94] reported a facile route for synthesizing ultrathin and flexible composite films based on rGOwrapped sulfur particles with the help of sodium alginate (SA) aqueous binder, which worked as a surfactant and an adhesive agent. The SA-glued electrode battery exhibited a high reversible capacity of 1341(818) mA h $g_{sul(ele)}^{-1}$ at 0.1C and retained its capacity 823(502) mA h $g_{sul(ele)}^{-1}$ at 0.5C after 100 cycles, which are more better compared to physically mixed rGO/S film. Therefore, in order to improve electrical conductivity and their mechanical stiffness, researchers made hybride by mixing SA with polyaniline and used as glue for rGO/Mn_3O_4/S nanocomposite particles electrode films prepared. They exhibited a high capacity of 1015(538) mA h $g_{sul(ele)}^{-1}$ at 5.0 A $_{gsul}^{-1}$ (B3.0C) and capacity retention of 71% after 500 cycles [95].

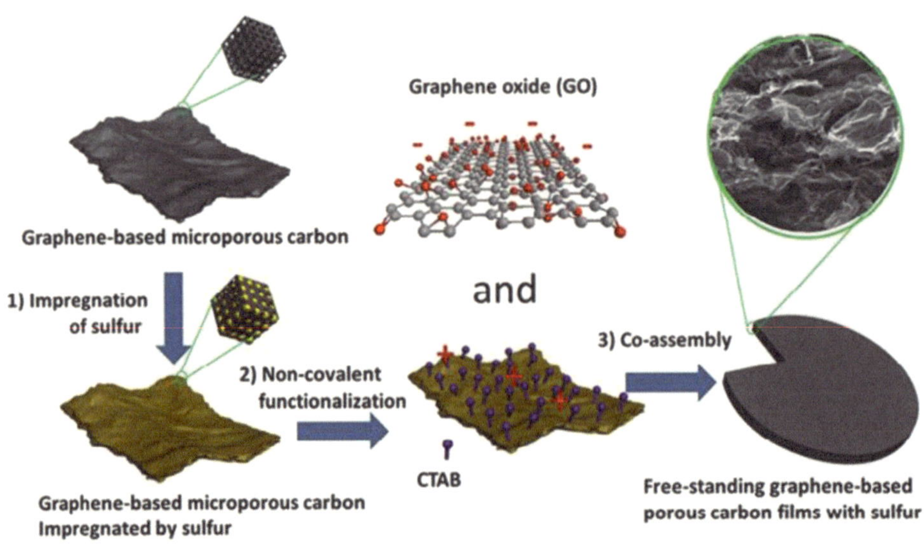

Figure 17.
Preparation of free-standing graphene-based porous carbon (GPC) films 1) impregnation of sulfur into the micropores of 2D graphene-based microporous carbon (GMC) sheets; 2) non-covalent functionalization of carbon (GMC)-sulfur sheets by CTAB. 3) assembly of positively charged GMC-sulfur sheets and negatively charged graphene oxide.Reprinted from ref. [93].

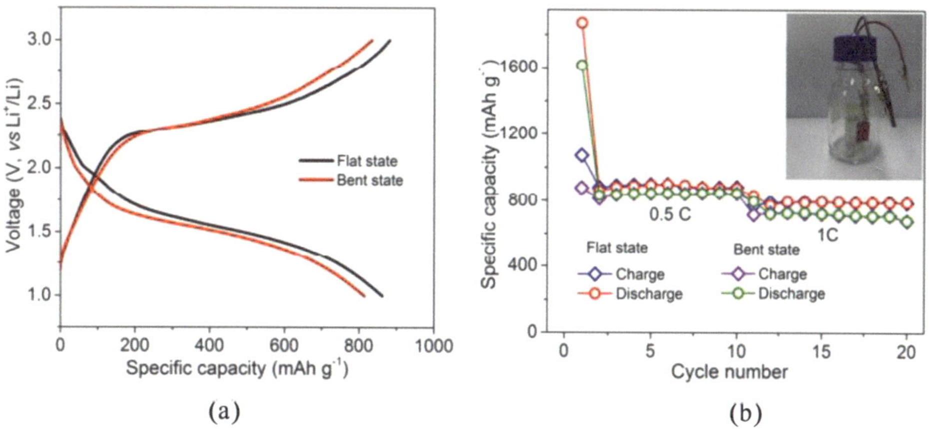

Figure 18.
(a) The second charge-discharge profiles of the GPC-sulfur cathode films at the bent and flat states at 0.5C. (b) the cycle performance for the GPC-sulfur cathode films at 0.5C and 1C, and inset showing that a bent cell is encapsulated in the glass bottle filled with argon. Reprinted from ref. [93].

2.7.2 Flexible Li: Se batteries

Selenium has several major merits for serving as cathode materials over to sulfur: (1) The magnitude higher electrical conductivity is an approximately 1024 times higher (2) More stable at room temperature, chain-like allotrope h-Se is more electro-active and more easily stabilized via spatial confinement (3) Selenium has more compatibility with conventional, cheap carbonate-based LIB electrolytes [96]. Therefore, selenium exhibits a better utilization rate, cyclic stability, and rate capability than sulfur. The volumetric specific capacity of h-Se is 3265 mA h cm^{-3} comparable to sulfur, 3461 mA h cm^{-3}, therefore it is more suitable for portable electronic devices and electrical vehicles due to its volume sensitive. Theoretically, the Li–Se battery utilizing h-Selenium as cathode lithium metal as anode, respectively at average voltage of 2.0 V, affords high gravimetric and volumetric energy densities of 1155 W h kg^{-1} and 2528 W h L^{-1}, respectively. Han et al. [68] introduced the mesoporous carbon nanoparticles (MCNs) with smaller size of 50 nm and favorable mesopore dominance, efficiently eliminated agglomeration in the bulk selenium. The electroactive selenium chains were stabilized in smaller micropores or mesopores, enabling high utilization and good cycling stability according to previous reports of Se–micro–/mesoporous carbon composite cathodes [84, 97]. The flexible Se/MCN–rGO cathodes demonstrated an ultrahigh selenium utilization of 97% at 0.1C, i.e., 655(406) mA h $g_{sel(ele)}^{-1}$. They exhibited good long cycling life with 89% capacity retention after 1300 cycles at 1.0C. This work is one of the most remarkable achievements for flexible Li–Se batteries by considering the high content of selenium. Similarly, Yu and Zhu's group prepared the composite PCNFs are represented as f-PCNFs, and they maintained good flexibility after selenization as shown in **Figure 19a–d** [98]. Very less crystalline selenium was present in PCNFs than in f-PCNFs leads to a remarkable improved capacity and initial Coulombic efficiency as shown in **Figure 19e**. The capacity and initial Coulombic efficiencies are 643/322 mA h $g_{sel/ele}^{-1}$ and 56.9% for Se@PCNFs at 0.05 A g_{sel}^{-1}, while 405/203 mA h $g_{sel/ele}^{-1}$ and 34.9% for Se@f-PCNFs. This is attributed due to the suppression of side reactions between free polyselenides produced from bulk selenium and carbonate electrolytes. Additionally, owing excellent encapsulation of selenium in the 1D conductive porous skeleton, flexible Se@PCNF cathode also exhibited non fading cycling performance with a capacity of 516(270) mA h $g_{sel(ele)}^{-1}$ retained after 900 cycles at 1.0 A g_{sel}^{-1} (B1.5C). The same electrospun PCNF–CNT also demonstrated in flexible Li–S batteries as like flexible selenium PCNF–CNT fabricated, battery exhibited reversible capacity of 638(223) mA h $g_{sel(ele)}^{-1}$ after 80 cycles at 0.05 A g_{sel}^{-1} (B0.074C) [99]. The utilization of conductive selenium (94%) was much higher than that of sulfur (38%) for the same PCNF–CNT conductive backbone demonstrated in cathodes.

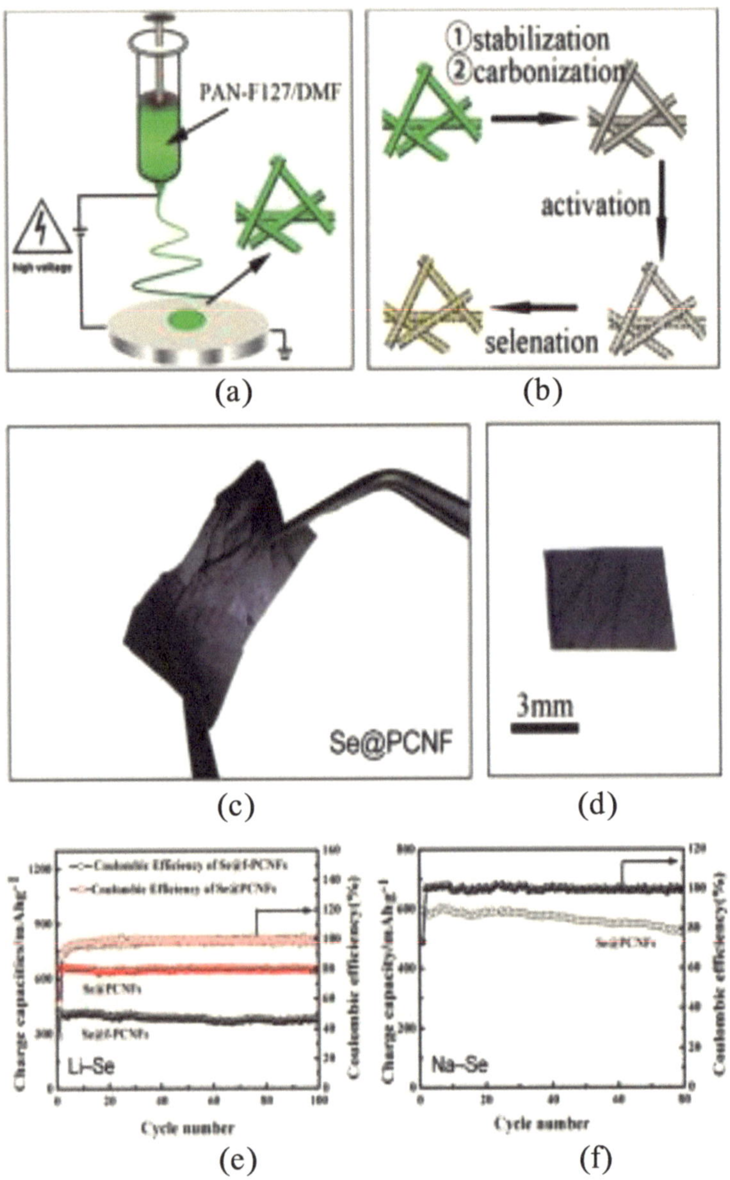

Figure 19.
Schematic representation of flexible selenium cathodes. (a)&(b) represents the synthesis of selenium (Se)@PCNF electrodes. (c) and d) picture of flexible Se@PCNF electrode. Cyclic performance of flexible Se@PCNF and Se@f-PCNF cathodes in (e) Li–Se and (f) Na–Se batteries at 0.05 a g_{sel}^{-1}*. [reprinted from ref. [98].*

2.7.3 Flexible Li–Te batteries

Tellurium is the last nonradioactive element in the chalcogen family, which exhibit highest electrical conductivity of 2.5 Scm^{-1} compared to all nonmetallic materials. Te shows low gravimetric specific capacity of 420 mA h g^{-1} due to its heavy atomic weight, it exhibited comparable volumetric specific capacity of 2621 mA h cm^{-3} to that of sulfur or selenium. Te is an electrically conducting active material required less carbon to prepare electrode. The decreasing weight of the carbon favors both volumetric and gravimetric specific capacities. A Li–Te battery exhibited theoretical gravimetric and volumetric energy densities of 682 W h kg^{-1} and 2078 W h L^{-1}, respectively with an estimated output voltage of 1.8 V. Li–Te battery was first demonstrated by Wang's group [100], in which tellurium/porous carbon composite cathode and a carbonate electrolyte as the components of battery. The Li–Te battery showed an average voltage of 1.5 V and a reversible capacity of 224 mA h g_{tel}^{-1} at 0.05 A g_{tel}^{-1}. They observed that 87% retention after 1000 cycles. Considering the relatively low voltage and promising volumetric capacity, Guo and coworkers demonstrated tellurium/carbon composites as anode materials for LIBs, indicating that extremely high tellurium utilization of 98% and a long-term cycling stability [101]. Particularly, Te is quite interest for flexible electrode materials due to its two most favorable features:

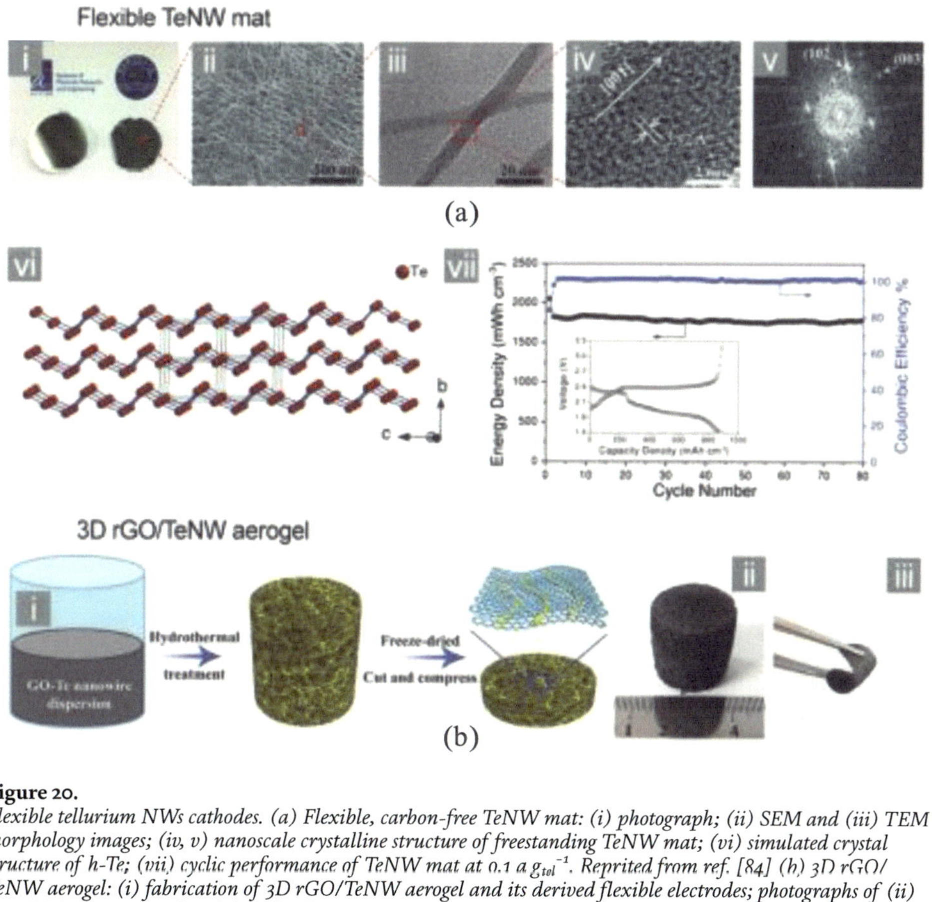

Figure 20.
Flexible tellurium NWs cathodes. (a) Flexible, carbon-free TeNW mat: (i) photograph; (ii) SEM and (iii) TEM morphology images; (iv, v) nanoscale crystalline structure of freestanding TeNW mat; (vi) simulated crystal structure of h-Te; (vii) cyclic performance of TeNW mat at 0.1 a g_{tel}^{-1}. Reprited from ref. [84] (b) 3D rGO/TeNW aerogel: (i) fabrication of 3D rGO/TeNW aerogel and its derived flexible electrodes; photographs of (ii) 3D rGO/TeNW aerogel and (iii) flexible 3D rGO/TeNW electrode. Reprinted from ref. [27, 103].

(1) high electrical conductivity compared to carbon and (2) The formation of 1D Te nanostructures along the c-axis, i.e., [001] due to its inherent chirality of helical Te chains in the h-Te crystal [102]. Hence, freestanding films consisting of ultralong Te NWs used directly used as an electrodes. Freestanding Te mat via vacuum filtration of Te NWs with a diameter of 7 nm grown in the [001] direction developed by Ding et al. as shown in **Figure 20**a-i–iii [104]. Such a high anisotropic 1D Te nanostructure exhibited fully Te zig-zag chains to lithium ions transport and a showed high electrical conductivity of 6.7 Scm^{-1} in the direction perpendicular to the c-axis as shown in **Figure 20**a(iv–vi). The flexible tellurium cathode comprises the Te NWs along with the new electrolyte exhibited a desirable capacity of B144 mA h $g_{tel}/_{ele-}{}^{1}$ at 0.1 A $g_{tel}{}^{-1}$ (0.24C). The volumetric energy density of 1800 W h L^{-1} observed after 80 cycles as shown in the **Figure 20**a-vii. Further, He and Chen's et al. [103] demonstrated a flexible tellurium cathode prepared from a 3D hierarchical aerogel with Te NWs wrapped homogeneously by rGO as shown **Figure 20**b. The synthetic method was adopted from the previous report on 3DCG–Li_2S. The rGO/Te NW electrode made of 63 wt% tellurium delivers high capacities of 418(263) and 174(110) mAh $g_{tel(ele)}{}^{-1}$ at 0.2 and 10C, respectively and excellent long-cyclic performance at a high rate of 1.0C. Therefore, as prepared flexible Te-NWs electrodes are quite attractive over the sulfur and selenium counterparts due to their distinguish features.

3. Conclusions

Al-S batteries prepared based on S/HPCK cathode delivered a capacity of 1027 mA h g^{-1} at 0.2 A g^{-1} for 50 cycles and exhibited excellent cyclic ability 405 mA h g^{-1} at 1 A g^{-1} for 700 cycles due to large porous structure with high surface area by adding of carbon powder. The Al-Se battery based on MCFs material with pore size of 7.1 nm exhibited a good capacity of 366 mA h g^{-1}. The chloroaluminate ion diffusivity greatly improved, which enhances the electrochemical behavior of Al-Se batteries. The rGO materials are introduced in the Al-Te batteries exhibited excellent cyclic ability and its initial capacity of 935.5 mA h g^{-1} and showed 467.5 mA h g^{-1} after 150 cycles with the Te loading of 70 wt% due to their excellent encapsulation.

The Interfacial layer of S@ spinel Ni-Co oxide double-shelled microspheres (NCO-HS) prepared under high sulfur loading exhibited minimum capacity fading rate of 0.045% per cycle over 800 cycles with high areal capacity of 6.3 mAh cm^{-2} and superior rate capability up to 5 C. As prepared capping layer of MTO-CNTs on surface of the sulfur cathode exhibited reversible capacity of 1212 mAh g^{-1} at the 0.2 C. The cycling stability of the device with the capping layer at 0.5 C rate still it retains a capacity of 577 mAh g^{-1} after 500 cycles with capacity decay rate of 0.07% per cycle, indicating a good cycling stability.

There is a gradual deterioration in all the compounds in the case of mixed cationic and anionic cyclic performance at a very low current density of 10 mA/g for 50 continuous cycles in terms of capacity fade. Cycling test of the 0.2 Fe substituted cathode done at 50 mA/g, reveals that the high rate cycling stability about 76% retention after 80 cycles. In the case of polysulfides on lithium deposition, the addition of Te exceeds 80% of its peak capacity for nearly 150 cycles and retains their cycling capacity without rapid drop until the electrolyte dryout nearly 300 cycles.

The developments of high-performance $Se_{1-x}S_x$ cathodes such as NC@SWCNTs@$Se_{0.2}S_{0.8}$ cathode exhibits good cycling stability (632 mA h g^{-1} at 0.2 A g^{-1} at 200 cycle) and high rate performance (415 mA h g^{-1} at 2 A g^{-1}) due to well-designed structure as well as optimized chemical composition with in carbonate-based electrolyte. Synthesized a nanoporous Co and N-co-doped carbon nanoreactor (C–Co–N) provide a high Te loading (77.2 wt%) provide ultrahigh capacity of 2615.2 mA h cm^{-3} and superior rate performance of 894.8 mA h cm^{-3} at 20C. Design structure and micro-environmental of Te-based nanoreactors could provide high electrochemical performance.

In the case of flexible batteries, in order to improve electrical conductivity and their mechanical stiffness, researchers made hybride by mixing SA with polyaniline and used as glue for rGO/Mn_3O_4/S nanocomposite particles electrode films, exhibited a high capacity of 1015(538) mA h $g_{sul(ele)}^{-1}$ at 5.0 A $_{gsul}^{-1}$ (B3.0C) and capacity retention of 71% after 500 cycles. The electrospun PCNF–CNT also demonstrated in flexible Li–S batteries as like flexible selenium PCNF–CNT fabricated, battery exhibited reversible capacity of 638(223) mA h $g_{sel(ele)}^{-1}$ after 80 cycles at 0.05 A g_{sel}^{-1} (B0.074C). The rGO/Te NW electrode made of 63 wt% tellurium delivers high capacities of 418(263) and 174(110) mAh $g_{tel(ele)}^{-1}$ at 0.2 and 10C, respectively and excellent long-cyclic performance at a high rate of 1.0C. Therefore, as prepared flexible Te-NWs electrodes are quite attractive over the sulfur and selenium counterparts due to their distinguish features.

Author details

Varishetty Madhu Mohan[1*], Madhavi Jonnalagadda[2] and VishnuBhotla Prasad[2]

1 Rajiv Gandhi University of knowledge Technologies, Kadapa, AP, India

2 Indian Institute of Science, Bangalore, India

*Address all correspondence to: madhuv1111@gmail.com; vmm@rguktong.ac.in

References

[1] Hao J, Xiong K, Zhou J, Rao AM, Wang X, Liu H, et al. Energy & Environmental Materials. 2021;**5**(1): 261-269

[2] Wang X-X, Denga N-P, Wei L-Y, Yang Q, Xiang H-Y, Wang M, et al. Chemistry, an Asian Journal. 2021;**16**(19):2852-2870

[3] Zhang Q, Cheng X, Wang C, Rao AM, Lu B. Energy & Environmental Science. 2021;**14**:965

[4] Liu Q, Rao AM, Han X, Lu B. Advancement of Science. 2021;**8**:2003639

[5] He S, Zhang D, Zhang X, Liu S-Q, Chu W-Q, Yu H-J. Advanced Energy Materials. 2021;**11**(29):2100769

[6] Zhang X, Jiao S, Tu J, Song W-L, Xiao X, Li S, et al. Energy & Environmental Science. 1918;**2019**:12

[7] Zhang Y, Liu S, Ji Y, Ma J, Yu H. Advanced Materials. 2018;**30**:1706310

[8] Boyjoo Y, Shi H-D, Tian Q, Liu S-M, Liang J, Wu Z-S, et al. Energy & Environmental Science. 2021;**14**:540575

[9] Huang C, Sun T, Shu H, Chen M, Liang Q, Zhou Y, et al. Electrochimica Acta. 2020;**334**:135658

[10] Saha S, Assat G, Sougrati MT, Foix D, Li H, Vergnet J, et al. Nature Energy. 2019;**4**(11):977-987

[11] Hansen CJ, Zak JJ, Martinolich AJ, Ko JS, Bashian NH, Kaboudvand F, et al. Journal of the American Chemical Society. 2020;**142**(14):6737-6749

[12] Flamary-Mespoulie F, Boulineau A, Martinez H, Suchomel MR, Delmas C, Pecquenard B, et al. Energy Storage Materials. 2020;**26**:213-222

[13] Nagarajan S, Hwang S-Y, Balasubramanian M, Thangavel NK, Reddy Arava LM. Journal of the American Chemical Society. 2021;**143**:15732-15744

[14] Li Z, Liu J, Huo X, Li J, Kang F, Appl ACS. Materials Interfaces. 2019;**11**:45709

[15] Lin S, Chen Y, Wang Y, Cai Z, Xiao J, Muhmood T, et al. Materials & Interfaces. 2021;**13**:9955-9964

[16] Zhang W, Li S, Wang L, Wang X, Xie J. Sustain. Energy & Fuels. 2020;**4**: 3588-3596

[17] Xu Q-T, Xue H-G, Guo S-P. Inorganic Chemistry Frontiers. 2019;**6**:1326-1340

[18] Du H, Feng S, Luo W, Zhou L, Mai L. Journal of Materials Science and Technology. 2020;**55**:1-15

[19] Sun J, Du Z, Liu Y, Ai W, Wang K, Wang T. Advanced Materials. 2021;**33**:2003845

[20] Jiao H, Tian D, Li S, Fu C, Jiao S, Appl ACS. Energy & Materials. 2018;**1**:4924

[21] Faegh E, Ng B, Hayman D, Mustain WE. Nature Energy. 2021;**6**:21

[22] Dong S, Yu D, Yang J, Jiang L, Wang J-W, Cheng L-W, et al. Advanced Materials. 2020;**32**(23):1908027

[23] Gupta A, Bhargav A, Manthiram A. Advanced Energy Materials. 2019;**9**: 1803096

[24] Shin H, Baek M, Gupta A, Char K, Manthiram A, Choi JW. Advanced Energy Materials. 2020;**10**:2001456

[25] Cheng X-B, Huang J-Q, Zhang Q. Journal of the Electrochemical Society. 2018;**165**:A6058-A6072

[26] Nanda S, Bhargav A, Jiang Z, Zhao X-H, Liu Y-Y, Manthiram A. Energy & Environmental Science. 2021;**14**: 5423-5432

[27] Peng H-J, Huang J-Q, Zhang Q. Chemical Society Reviews. 2017;**46**:5237-5288

[28] Chung SH, Manthiram A. Advanced Materials. 2019;**31**:1901125

[29] Ding H, Zhou J, Rao AM, Lu B. National Science Review. 2021;**8**(9):nwaa276. DOI: 10.1093/nsr/nwaa276

[30] Guo Y, Jin H, Qi Z, Hu Z, Ji H, Wan L-J. Advanced Functional Materials. 2019;**29**:1807676

[31] Zhang X, Wang B, He S, Liu S, Tang M, Yu H. Journal of Materials Chemistry A. 2021;**9**:8966

[32] Kong Y, Nanjundan AK, Liu Y, Song H, Huang X, Yu C. Small. 2019;**15**: 1904310

[33] Zhang X, Wang M, Tu J, Jiao S. Journal of Energy Chemistry. 2021;**57**:378

[34] Li F, Liu Q, Hu J, Feng Y, He P, Ma J. Nanoscale. 2019;**11**:15418-15439

[35] Liu T, Hu H, Ding X, Yuan H, Jin C, Nai J, et al. Energy Storage Materials. 2020;**30**:346-366

[36] Park JS, Kim JK, Hong JH, Cho JS, Park SK, Kang YC. Nanoscale. 2019;**11**: 19012-19057

[37] Cao B, Huang J, Mo Y, Xu C, Chen Y, Fang H, et al. Materials & Interfaces. 2019;**11**:14035-14043

[38] Liu Y, Kou W, Li X, Huang C, Shui R, He G. Small. 2019;**15**:1902431

[39] Zhao C, Ge F, Shi L, Wang H, Liu J, Zhang J, et al. International Journal of Hydrogen Energy. 2019;**44**:30478-30485

[40] Li F, Tao J, Zou Z, Li C, Hou Z, Zhao J. ChemSusChem. 2020;**13**: 2761-2768

[41] Li N, Chen Z, Chen F, Hu G, Wang S, Sun Z, et al. Carbon. 2019;**143**:523-530

[42] Li W, Chen Z, Wang D, Gong Z, Mao C, Liu J, et al. Journal of Power Sources. 2019;**435**:226778

[43] Mussa Y, Arsalan M, Alsharaeh E. Energy & Fuels. 2021;**35**:8365-8377

[44] Ma L, Zhu G, Zhang W, Zhao P, Hu Y, Wang Y, et al. Nano Research. 2018;**11**:6436-6446

[45] Du P, Wei W, Dong Y, Liu D, Wang Q, Peng Y, et al. Nanoscale. 2019;**11**:10097-10105

[46] Huang Y, Field R, Chen Q, Peng Y, Walczak MS, Zhao H, et al. Communications & Chemistry. 2019;**2**:138

[47] Xiang Y, Wang Z, Qiu W, Guo Z, Liu D, Qu D, et al. Journal of Membrane Science. 2018;**563**:380-387

[48] Shi H, Sun Z, Lv W, Xiao S, Yang H, Shi Y, et al. Journal of Energy Chemistry. 2020;**45**:135-141

[49] Wang N, Chen B, Qin K, Liu E, Shi C, He C, et al. Nano Energy. 2019;**60**: 332-339

[50] Li Y, Wang W, Liu X, Mao E, Wang M, Li G, et al. Energy Storage Materials. 2019;**23**:261-268

[51] Wang J, Yang G, Chen J, Liu Y, Wang Y, Lao CY, et al. Advanced Energy Materials. 2019;**9**:1902001

[52] Yan X, Zhang H, Huang M, Qu M, Wei Z. ChemSusChem. 2019;**12**: 2263-2270

[53] Hu Y, Zhu X, Wang L. ChemSusChem. 2020;**13**:1366-1378

[54] Gao TJ, Xu DP, Yu ZH, Huang ZH, Cheng J, Yang Y. Journal of Alloys and Compounds. 2021;**865**:11

[55] Chen YT, Abbas SA, Kaisar N, Wu SH, Chen HA, Boopathi KM, et al. Materials & Interfaces. 2019;**11**:2060-2070

[56] Martinolich AJ, Zak JJ, Agyeman-Budu DN, Kim SS, Bashian NH, Irshad A, et al. Chemistry of Materials. 2021;**33**(1):378-391

[57] Ramakrishnan S, Park B, Wu J, Yang W, McCloskey BD. Journal of the American Chemical Society. 2020;**142**(18):8522-8531

[58] Zhu Z, Yu D, Yang Y, Su C, Huang Y, Dong Y, et al. Nature Energy. 2019;**4**(12):1049-1058

[59] Yan P, Zheng J, Tang Z-K, Devaraj A, Chen G, Amine K, et al. Nature Nanotechnology. 2019;**14**(6):602-608

[60] Augustyn V, Come J, Lowe MA, Kim JW, Taberna PL, Tolbert SH, et al. Nature Materials. 2013;**12**:518-522

[61] Wujcik KH, Wang DR, Raghunathan A, Drake M, Pascal TA, Prendergast D, et al. Journal of Physical Chemistry C. 2016;**120**:18403-18410

[62] Zhang G, Peng H, Zhao C, Chen X, Zhao L, Li P, et al. Angewandte Chemie, International Edition. 2018;**57**:16732-16736

[63] Baek M, Shin H, Char K, Choi JW. Advanced Materials. 2020;**32**:2005022

[64] Nanda S, Manthiram A. Energy & Environmental Science. 2020;**13**: 2501-2514

[65] Bhargav A, He J, Gupta A, Manthiram A. Joule. 2020;**4**:285-291

[66] Cheng X-B, Yan C, Huang J-Q, Li P, Zhu L, Zhao L, et al. Energy Storage Mater. 2017;**6**:18-25

[67] Chen S, Niu C, Lee H, Li Q, Yu L, Xu W, et al. Joule. 2019;**3**:1094-1105

[68] Han K, Liu Z, Ye HQ, Dai F. Journal of Power Sources. 2014;**263**:85-89

[69] Gupta MK, Singh B, Goel P, Mittal R, Rols S, Chaplot SL. Physical Review B: Condensed Matter and Materials Physics. 2019;**99**:224304

[70] Zhang L, Ling M, Feng J, Mai L, Liu G, Guo J. Energy Storage Materials. 2018;**11**:24-29

[71] Nanda S, Bhargav A, Manthiram A. Joule. 2020;**4**:1121-1135

[72] Pearson RG. Journal of Chemical Education. 1968;**45**:581-587

[73] Sahu G, Lin Z, Li J, Liu Z, Dudney N, Liang C. Energy & Environmental Science. 2014;**7**:1053-1058

[74] Wang Y, Lu X, Zheng C, Liu X, Chen Z, Yang W, et al. Angewandte Chemie. 2019;**131**:7755-7759

[75] Lau J, DeBlock RH, Butts DM, Ashby DS, Choi CS, Dunn BS. Advanced Energy Materials. 2018;**8**:1800933

[76] Lu C-W, Fang R-Y, Wang K, Xiao Z, Kumar GG, Gan Y-P, et al. Frontiers in Chemistry. 2021;**9**:738977

[77] Zhu T, Pang Y, Wang Y, Wang C, Xia Y. Electrochimica Acta. 2018;**281**:789-795

[78] Guo B, Yang T, Du W, Ma Q, Zhang L-Z, Bao S-J. Journal of Materials Chemistry A. 2019;7:12276-12282

[79] Li X, Liang J, Zhang K, Hou Z, Zhang W, Zhu Y. Energy & Environmental Science. 2015;**8**:3181-3186

[80] Luo C, Zhu Y, Wen Y, Wang J, Wang C. Advanced Functional Materials. 2014;**2014**(24):4082-4089

[81] Li Z, Zhang J, Lu Y, Lou XW. Science Advances. 2018;**4**:1687

[82] Li Z, Zhang J, Wu HB, Lou XWD. Advances Energy Materials. 2017;7: 1700281

[83] Yao Y, Zeng L, Hu S, Jiang Y, Yuan B, Yu Y. Small. 2017;**13**:1603513

[84] Luo C, Xu Y, Zhu Y, Liu Y, Zheng S, Liu Y, et al. ACS Nano. 2013;7:8003-8010

[85] Li Z, Yuan L, Yi Z, Liu Y, Huang Y. Nano Energy. 2014;**9**:229-236

[86] Tian H, Tian H, Wang S, Chen S, Zhang F, Song L, et al. Nature Communications. 2020;**11**:5025

[87] Li Y, Zhang Y, Xu Q, Hu L, Shen B, Liu H, et al. ChemSusChem. 2019;**12**: 1196-1202

[88] Xu J, Ma J, Fan Q, Guo S, Dou S. Advanced Materials. 2017;**29**:1606454

[89] Zeng L-C, Li W-H, Jiang Y, Yu Y. Rare Metals. 2017;**36**:339-364

[90] He J, Lv W, Chen Y, Wen K, Xu C, Zhang W, et al. ACS Nano. 2017;**11**: 8144-8152

[91] Zhai PY, Huang JQ, Zhu L, Shi JL, Zhu W, Zhang Q. Carbon. 2017;**111**:493-501

[92] Shi JL, Peng HJ, Zhu L, Zhu WC, Zhang Q. Carbon. 2015;**92**:96-105

[93] Wu C, Fu LJ, Maier J, Yu Y. Journal of Materials Chemistry A. 2015;**3**: 9438-9445

[94] Ni W, Cheng JL, Li XD, Guan Q, Qu GX, Wang ZY, et al. RSC Advances. 2016;**6**:9320-9327

[95] Ghosh A, Manjunatha R, Kumar R, Mitra S, Appl ACS. Materials & Interfaces. 2016;**8**:33775-33785

[96] Yang CP, Yin YX, Guo YG. Journal of Physical Chemistry Letters. 2015;**6**:256-266

[97] Yang CP, Xin S, Yin YX, Ye H, Zhang J, Guo YG. Angewandte Chemie, International Edition. 2013;**52**: 8363-8367

[98] Zeng LC, Zeng WC, Jiang Y, Wei X, Li WH, Yang CL, et al. Advanced Energy Materials. 2015;**5**:1401377

[99] Zeng LC, Wei X, Wang JQ, Jiang Y, Li WH, Yu Y. Journal of Power Sources. 2015;**281**:461-469

[100] Liu Y, Wang JW, Xu YH, Zhu YJ, Bigio D, Wang CS. Journal of Materials Chemistry A. 2014;**2**:12201-12207

[101] Zhang J, Yin YX, You Y, Yan Y, Guo YG. Energy Technology. 2014;**2**:757-762

[102] He Z, Yang Y, Liu JW, Yu SH. Chemical Society Reviews. 2017;**46**:2732-2753

[103] He JR, Chen YF, Lv WQ, Wen KC, Wang ZG, Zhang WL, et al. ACS Nano. 2016;**10**:8837-8842

[104] Ding N, Chen SF, Geng DS, Chien SW, An T, Hor TSA, et al. Advanced Energy Materials. 2015;**5**:1401999

Section 3

Nanoscale Devices

Chapter 7

Temperature-Dependent Evaluation of Charge Carriers and Terahertz Generation in Bismuth and Antimony-Based Chalcogenides

Prince Sharma, Veerpal Singh Awana and Mahesh Kumar

Abstract

Bismuth and antimony-based chalcogenides have been extensively publicized in recent years owing to their intrinsic characteristics and inherent topological character. Such a system contains Bi_2Se_3, Bi_2Te_3, Sb_2Te_3, etc. The single crystalline facets of these samples were discovered to have a generation of ~2 THz while having a giant magneto-resistance of around ~300%. These inherent and dynamical features of the system make it resilient for several applications in optoelectronics and spintronics. The temperature-dependent assessment of conductivity, terahertz generation, and charge carrier dynamics aids in understanding the fundamental phenomena in the carrier mechanism of the chalcogenides. This chapter contains the essential fundamental knowledge of the single crystal chalcogenides via charge carrier & phonon dynamics and their response in the terahertz frequency domain.

Keywords: topological material, temperature-dependent carrier dynamics, terahertz, transport properties

1. Introduction

The quantum interaction of charge carriers with external and internal forces in different materials remains an open question for the condensed matter community. The enigma began with Edwin Hall's discovery of the classical Hall Effect in 1879 [1]. The Hall effect shows a voltage difference formed by injecting steady current and magnetic fields across a conductor or semiconductor. Due to the current and external magnetic fields interplay, voltage generation occurs. This voltage difference is due to charge confinement. In the 1980s, Von Klitzing discovered the quantum hall effect (QHE) in a two-dimensional system [2]. Due to the intense magnetic fields, charge carriers are constrained into two dimensions and exhibit topologically ordered states. The edge states at the surface cause current to flow at the superiorities, and these states are formed as a result of the high external magnetic field. These geometrical states give birth to a new phase of matter known as topological insulators (TI) [3]. Charles Kane and Eugene Mele anticipated the development of TI in 2005 [3–8]. It is very similar to QHE, except that no external magnetic field is necessary since the inherent characteristics of materials generate the magnetic field. The spin-orbit interaction generates this magnetic field. TI is unusual because it is insulating in bulk yet conducts at the surface. The electronic wave function of a charge carrier is dependent on its shape, which varies from bulk to surface. As a result, it is referred to as a topological insulator. The SS renders the system impervious to non-magnetic doping due to protected SS. These SS are protected by time-reversal symmetry (TRS). This small property enables many unique applications due to the impervious topological states to non-magnetic disruption and their dissipation-free spin current transit. Spintronic, thermoelectric, magnetic memory storage, magnetoelectric devices, next-generation batteries, THz generators, transistors, photodetectors, and sensor applications are only a few of these applications, which is possible in these TI [3–13].

This article is focused on topological insulators, providing an overview of the topological phases and states found in TIs. Additionally, the dynamics of the carrier and phonon scattering are also discussed. TI's surface and bulk states are probed using various optical methods. The ultrafast laser pulse is employed in particular to characterize the functional characteristics of Fermions in TI. These pulses have also been used to examine the phonon vibration mode. Finally, it establishes the existence of the coherent optical phonon (COP) and coherent acoustic phonon (CAP) modes. The temperature-dependent evolution of these modes has also been examined as these phonon vibrations progress with the charge carrier dynamics. Transient Reflectance Ultrafast Spectroscopy (TRUS) was utilized to explore the non-conventional conducting surface charge carriers. A femtosecond pump beam was employed to stimulate the sample. The material was probed with a wide beam ranging from visible to near-infrared. TRUS measurements aid in determining the charge carrier dynamics and the capacity of terahertz production. Additionally, investigating carrier and phonon dynamics in a temperature-dependent manner aids in the understanding of crucial transitions associated with surface states.

2. Transient reflectance ultrafast spectroscopy (TRUS)

J.Qi. et al. performed the first TRUS measurements on Bi_2Se_3 crystals [14], where they demonstrated the time-resolved behavior of the crystal at room temperature. Three distinct relaxations of carriers induced by photons were identified in this experiment. The first two are phonon interactions between excited charge carriers of coherent optical and acoustic phonon interactions. In contrast, the third is a negative amplitude process generated by ultrafast carrier trapping of selenium. They also observed the frequency of 2.13 THz from their optical phonon oscillations. They also concluded that the atmosphere impacts these charge carriers and phonons since air promotes band bending, which causes an elevation in the Fermi level [14]. Nardeep Kumar et al. [15] utilized the ultrafast pulse to analyze the carrier's behavior by pump-probe spectroscopy. The Bi_2Se_3 TI exhibits two distinct oscillations, with the rapid oscillations occurring at 2.167 THz and the slow oscillations occurring at 0.033 THz. Coherent optical and acoustic phonons are responsible for this frequency [15]. In these systems, the terahertz frequency is not dependent on the size of the laser spot, and the ambipolar carrier diffusion coefficient is also determined to be 500 cm^2/s [15]. Nardeep et al. have shown that the COP-induced terahertz frequency production has an uneven dependency.

The ultrafast carrier dynamics were further investigated in the limited thickness of the TI, using pump-probe spectroscopy in thin films by Yuri D. Glinka et al. [16]. The thicker films have similar relaxation lifetimes to bulk crystals, while the thinner films exhibit fastened relaxation of excited charge carriers. The longevity of SS and bulk states might be related to their contribution. The shorter carrier lifetimes in thin films are associated with reducing bulk contributions of carriers in dominant surface states. A resonance-like property is also seen in 10 nm films, but the study does not provide conclusive evidence for how the property is generated. Yuri D. Glinka et al. also investigated the thin films of Bi_2Se_3 (6–40 nm), in which they confirmed the presence of radiative and non-radiative processes and described resonance phenomena at 10 nm films in terms of these processes [17]. This article establishes unequivocally that bismuth selenide contains a second SS. After the detection of the second SS, the primary trend observed in TRUS is the presence of three distinct processes: (a) electron–electron and electron longitudinal optical phonons in the 1–8 ps range, (b) a metastable bulk conduction band that continuously feeds charge carriers to the second SS for approximately 10 ps, and (c) a quasi-equilibrium carrier population. The thin layer is activated by 1.51 eV photons that excite carriers from the bulk conduction band (BCB) to the second surface states (SS). From the second surface to the first SS, carriers undergo intra-band and inter-band relaxation. The relaxations occur between the second SS and BCB and first SS, indicating that the ultimate recombination happened in these states. The relaxation of the excitation charge carriers results in forming a new Fermi level that is displaced away from the original Fermi energy. This relocation is a result of carrier localization. Again, the resonance phenomenon in a 10 nm film is found in this pump-probe spectroscopy, which is explained by the depletion of electrons caused by the connection between two film surfaces. The confinement of these 3D electrons is due to the existence of surface defects caused by selenium vacancies; however, they grew the 10 nm film many times and obtained this resonance in each of these films. This repeatability lends credence to the selenium vacancy theory. This resonance effect in 10 nm TI is unknown at the moment However, recent work by Glinka et al.; explains the discrepancy by suggesting several surface state amendments [18].

3. Terahertz generation at room temperature

Because of the expansion of the communication systems, information processing, and transmission fields are the most susceptible [19–22]. In terms of the frequency range, the lower Terahertz range may benefit air transmission, while the higher frequency range enables faster signal transfer and can be used to create ultra-high bandwidth data links. A fundamental component of numerous fields, including information processing and transmission, security screening, and biomedical applications, are emerging from Terahertz research [19, 21, 23, 24]. The future applications of Terahertz radiations have prompted an influx of scientific and technical research into creating Terahertz pulses for bio-material imaging, ultrafast dynamics, and nonlinear Terahertz optics [21, 25], among other things. This frequency may be generated in various ways, one of which is by using an ultrafast and ultrashort laser pulse. The interaction of short laser pulses with various objects makes it more challenging to produce adjustable, compact, coherent, and high-magnitude Terahertz radiation sources. The two-color filamentation in gases is a simpler way for a generation since it can scale the magnitude to incredible levels of interest and complexity.

However, the scaling was limited by the laser pulse energy constraints, and as a result, the search for alternate target materials was essential for these applications to succeed. It is necessary to search in a varied and uncluttered study region to locate the target, which exhibits Terahertz generations of high energies. The different crystals are the subject of this study due to their crystalline nature, as they can be used for large-scale broadband and high-frequency terahertz generation. Accordingly, phonon dynamics of topological insulators single crystals provide an attractive benefit in terahertz generator performance.

The terahertz frequency range of 0.03 to 5.2 THz is created by a single crystal of Bi_2Se_3 [14, 26–38]. In summary, phonons, which are the COP in bismuth selenide, are responsible for generating the terahertz frequency, as mentioned in the TRUS section above. The terahertz spectroscopy directly validates the 2 THz frequency generation, while the TRUS was the initial method by which these oscillations were originally seen [14, 18, 26, 30, 32, 34, 36, 37, 39–53] and was used to confirm the frequency of the 2 THz frequency. In addition, Prince Sharma et al. in 2020 contrasted assessing the frequency determined through TRUS, revealing the COP oscillations [34, 48] to evaluate the frequency from the kinetic spectrum. **Figure 1** depicts oscillations obtained from the kinetics of a flake cleaved out of the large single crystal of Bi_2Se_3 that was recorded at the CSIR-NPL. The FFD (filtering the high-frequency component and fitting the data) analysis investigates the generated frequency. These oscillations are removed from the kinetic data profile of charge carriers by applying a high pass filter in the 1 to 10 ps period. The data that has been filtered out is fitted with the sinusoidal damped function to determine the frequency associated with these oscillations. The frequency associated with them is determined to be ~2.42 terahertz (THz) from FFD analysis.

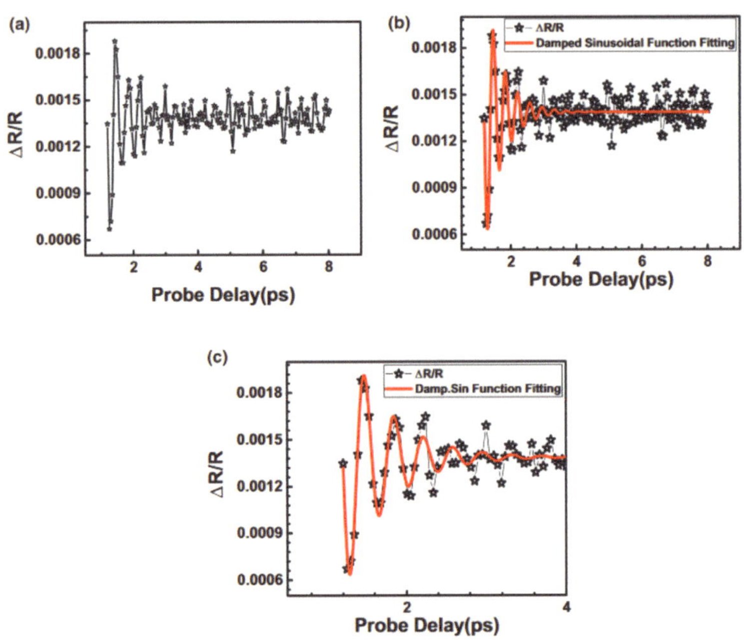

Figure 1.
(a) Illustration of optical phonon oscillations by filtering with a high-pass Fourier filter with a cutoff frequency of 2.32 THz. (b) The optical phonons vibrations are fitted with a damped sinusoidal function. (c) The amount of experimental fitting data has skyrocketed. Reprinted from ref. [48], with permission of springer nature.

A significant expansion on the assessment of terahertz production is also carried out, where the terahertz frequency may be changed just by altering the interaction of electromagnetic radiation in a single crystal. First and foremost, the probe energy is monitored while the excitation energy is maintained consistently. We detected strong oscillations at 1100 nm with a frequency of 2.42 THz in this system. A continuous probe at 1100 nm is then monitored by changing the excitation wavelength, which is the second step. In **Figure 2**, the tunable nature of the single-crystalline flake of bismuth selenide [34] is shown in detail. It is discovered that the frequency of phonon vibrations may be controlled by tuning the carriers in the crystal by appropriate doping, as well [45, 54]. Consequently, the adjustable nature of terahertz in single crystals might be beneficial in the growth of optoelectronics and Terahertz applications.

Although the primary emphasis is on the COP dynamics, Yuri D. Glinka et al. return to the distinct relaxations, namely, CAP [55]. The paper noted an increase in CAP frequency from 35 to 70 GHz. When the thickness of TI was reduced, the interaction of two processes caused this frequency difference. As the film thickness goes below 15 nm, lamb wave excitations (elastic waves whose propagation is plane to plane) become visible. Above this critical thickness, the system operates in the bulk acoustic resonator mode (indirect inter-surface coupling). Apart from performing TRUS measurements at ambient temperature, Yi-Ping Lai et al. conducted a temperature-dependent examination between 11–294 K [44]. To begin, they summarized the physical processes occurring at various time scales predicted by pump-probe experiments as fast oscillations (10^{12} Hz–COP), slow oscillations (10^{10} Hz–CAP), non-oscillatory signal (10^{11} Hz – electrons and incoherent phonons), and constant residual (10^{11} Hz – slow electron dynamics).

The temperature-dependent evolution of the COP revealed that when the sample approaches room temperature, the optical phonon frequency reduces from 2.25 to 2.17 THz [44]. This investigation also determines the electron–phonon coupling constant, demonstrating that the observed signal dominates the bulk. Additionally, Sung Kim et al. [37] studied the compatibility of the thin films with various polycrystalline and crystalline substrates and the resonance impact of the same at various thicknesses ranging from 3 to 30 nm. At 2.1 and 5.2 THz, the intensity of the differential reflectance (DR) and two distinct phonon modes exhibited some resonant behavior within a specific range of a crucial number of quintuple layers (QL). This resonant effect is substrate-independent, and the crucial QL value is between 6 and 9 QLs. Apart

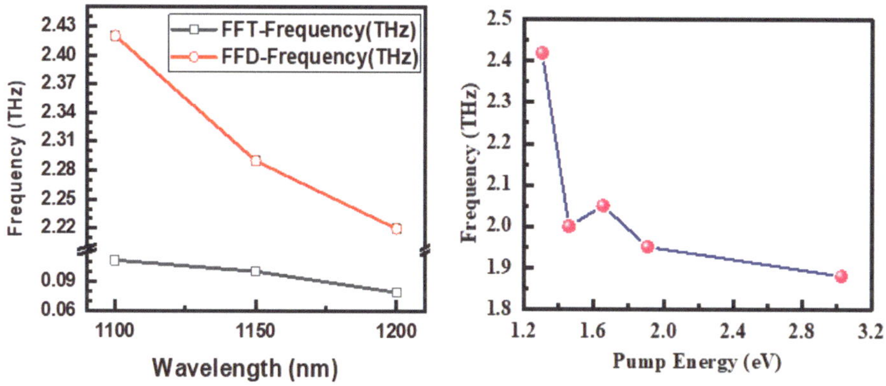

Figure 2.
The terahertz frequency is dependent on the probe wavelength and pumps excitation energy. Reprinted from ref. [34], with permission of springer nature.

from this resonant effect, Jianbo Hu et al. [49] regulated the strong Raman mode of bismuth selenide using a two-pump laser in TRUS studies. They demonstrated phonon chirping due to the two-pump arrangement, which confirms the earlier thesis of carrier-lattice coupling. These two pumps effectively tune the amplitude, but the frequency stays constant.

Pump-probe spectroscopy is used to investigate single excitations with a specific energy. However, Giriraj Jnawali et al. employed the mid-infrared femtosecond TRUS on a nanoflake of TI for the first time [50]. The energy range is 0.3–1.2 eV. This range includes both the original BCB and the extended BCB. As a result, the paper modeled the ultrafast photoexcited carriers and holes across the BCB. Theoretical modeling of 10 K DR indicates a strong probability of the Fermi level being present much above the lowest conduction band. These experiments established a firm knowledge of the carriers inside TI's BCB, establishing a direct connection between carriers and Dirac SS [50].

Nonetheless, the TRUS provides no evidence for spin-polarized charges in SS through these broad energy probes. M.C. Wang investigated spin-dependent transitions that occurred solely in topologically protected SS using time-resolved Kerr rotation measurements [41]. Transitions between topologically protected SS may induce net spin polarization due to the presence of spin-protected carriers inside these SS. However, the study demonstrates that a circularly polarized pump may create net polarized spin, but the transitions responsible for this polarized spin are not just between the two SS but also between the first SS and the second BCB located near the second SS [41]. While the TRUS predicts just the energy levels of the different bulk conduction bands at ambient temperature. The surface state-related transitions are readily seen when investigating the temperature-dependent dynamics.

4. Temperature-dependent charge and phonon dynamics in Bi_2Se_3, Sb_2Te_3, and Bi_2Te_3

4.1 Bi_2Se_3

TRUS is used to investigate the exotic topological quantum characteristics of Bi_2Se_3, Sb_2Te_3, and Bi_2Te_3. This particular regime probing aids in comprehending the TI's enigmatic behavior. Considering the bismuth selenide, the PL emission indicates a significant 2 eV optical transition [56] caused by the state bunching effect [53]. The inert nature of these transitions is explained using density functional theory (DFT) calculations on the band structure and Kramer's Kroning method on reflectance data from crystal flake. Additionally, TRUS measurements are performed with a variety of pump excitation energies (3.02, 2.61, 1.91, and 1.4 eV) to obtain a spectrum in the VIS–NIR region (2.58–0.77 eV) [53]. These wide regime experiments on carrier dynamics demonstrate that the Moss–Burstein and shielding effects exist in bismuth selenide. Additionally, these studies demonstrate a variety of relaxation mechanisms, including thermalization of hot carriers, COP and CAP relaxation, and recombination.

All of the phenomena mentioned above are observed at room temperature, and there is a strong need to explore the TIs at low temperatures to assess surface state-induced transitions. Essentially, there are two distinct ways for probing surface states or surface-related transitions. We already know that TIs are conducted at the surface while insulating bulk. Thus, the surface states of TIs are located on the uppermost layer. Therefore, in order to explore these surface states, the sample thickness should be reduced to the ultrathin regime [17, 18, 55]. The sole disadvantage of this strategy is the complexity of the systems, required to develop this kind of ultrathin film. Alternatively, the low-temperature technique is used to investigate the temperature-dependent dynamics of charge carriers and phonons between 5 and 300 K. The perceptions of observing the surface states or surface states associated with transitions at low temperatures may be explained by performing a magneto-resistance analysis using the HLN equation. The literature demonstrates that at extremely low temperatures and magnetic fields, the surface states of TIs prevail over the bulk states [57–60]. Thus, it may be interesting to examine the charge and phonon dynamics of TIs at low temperatures in order to see surface states or surface state-related transitions.

The TRUS is carried out at low temperatures on micro flakes of single-crystalline bismuth selenide to observe the associated transitions in the temperature range of 5–300 K [53]. **Figure 3a** illustrates the DR spectra of crystalline flake across a wide range of NIR wavelengths at various temperatures. The wide DR signal demonstrates TI's ability to exhibit a broad range of optically allowed transitions, distinguished as 0.7, 1.1, and 1.4 eV. DFT band structure calculations anticipated that these specific transitions are permissible in bismuth selenide [53]. As seen in **Figure 3a**, a DR signal is denoted by a B peak that is only present above 200 K. At low temperatures, this specific transition is suppressed, indicating that it is associated with certain phonon-assigned carriers. Essentially, these carriers have initial energy of below 200 K, caused by thermalization at high temperatures.

Additionally, when the temperature decreases below 200 K, a tiny DR signal of 1.2 eV is formed at 100 K. The level of these DR signals increases when the temperature is lowered and becomes more noticeable below 5 K. These DR signals correspond to the same as the second strategy of exploring surface states associated transitions mentioned above. There are two possible explanations for this transition: defect-induced peak and surface state-related transition. If we consider the first option of

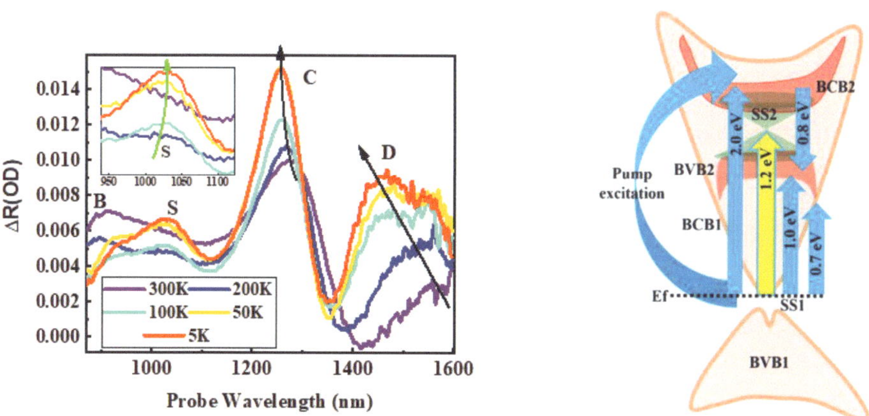

Figure 3.
(a) illustrates the differential reflectance at 750 fs of a single crystal throughout the whole NIR range (800–1600 nm) at temperatures ranging from 5–300 K. It demonstrates the presence of a blue shift with the temperature that happened as a result of thermal fluctuations being suppressed. A decrease in DR at 1000 nm at low temperature coincides with the surface state transition, confirming the shift to the second surface state. (b) Theoretical transition model in which BCBs and BVBs are drawn to resemble bands in the same way as DFT calculations are performed on an ideal system. It is a DFT-based model, and TRUS predicted a variety of OBTs. These OBTs have a threshold voltage of 0.7, 1.0, 1.3, and 2.0 eV and stimulated emission of 0.8 eV. Additionally, using low temperature TRUS verifies the occurrence of a second surface condition. Reprinted from ref. [53], with permission of Elsevier.

a defect-induced peak, then this kind of transition has a common property, i.e., the carrier relaxation lifetime must be very short. However, the lifetimes in this situation are in the picosecond range, which eliminates the likelihood of a defect-induced peak [61]. Additionally, the kinetic profile of the same does not alter with temperature, corroborating the preceding explanation. Thus, one thing is evident that this specific DR signal is not the result of a defect.

After establishing that the DR signal at 1.2 eV is not attributable to defects, it is not incorrect to assert that the signal is predicted to be due to a surface state-related transition. DFT calculations are used to determine the band structure using the effective SOC inclusion to resolve this particular uncertainty. After learning about the band structure, it was relatively simple to formulate the 1.2 eV transition. This transition occurs when the carrier is excited from its ground state to its second surface state. Additionally, the kinetic profile during this transition is fitted using three lifetimes, and the fitting of this kinetic profile indicates that carriers relax in the picosecond time domain when they migrate from this energy level to a lower energy level. In the case of noble metals, when the relaxation time is in the picosecond range, it is widely known that the surface states reveal metallic nature. However, in our case, too, the surface states exhibit metallic properties. As a result, the relaxation lifetime suggests the occurrence of a surface state-related transition in TI [53].

Band structure calculations utilizing DFT in conjunction with actual pump-probe spectroscopy may be used to predict various optically allowed transitions in bismuth selenide. First surface state below Fermi level (SS1), second surface state above second valence band (SS2), first bulk valence band (BVB1), second bulk valence band (BVB2), first bulk conduction band (BCB1), and second bulk conduction band (BCB2) of TI are shown in **Figure 3b**. The temperature-dependent TRUS response is shown in **Figure 3a**. The most prominent optical transitions in TRUS are ~0.7, ~1.0, ~1.4, and ~ 2.0 eV. Charge carriers are stimulated to the second bulk valance band, exhibiting ~0.7 and ~ 1.0 eV DR signal peaks, whereas the ~2.0 eV transition occurs in the second bulk conduction band. Additionally, the low-temperature investigation reveals the existence of a DR peak of about 1.2 eV, which corresponds to the transition to the second surface state. Thus, the temperature-dependent study of charge carrier dynamics enables the investigation of various bands and surface-related transitions.

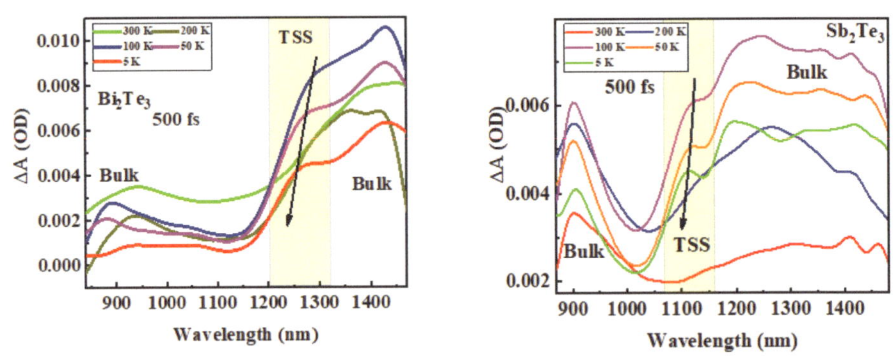

Figure 4.
The temperature-dependent differential reflectance (DR) at 500 fs is analyzed using micro-flakes stimulated at 3.02 eV and probed in the 1.55–0.77 eV region of (a) Bi2Te3 and (b) Sb2Te3, respectively. The peaks highlighted in the TSS region of both TIs exhibit a blue shift when the temperature decreases, which is attributable to a reduction in the thermalization process. Reprinted from ref. [62], with permission of Elsevier.

4.2 Bi_2Te_3 & Sb_2Te_3

Temperature-dependent TRUS measurements are also performed in the near-infrared region on micro flakes of bismuth telluride. At room temperature, many optically allowed transitions are detected in TIs. Additionally, it was discovered that the DR response below 100 K exhibits a noticeable transition that is connected to the surface state of bismuth telluride and antimony telluride. At 500 fs, **Figure 4** depicts the temperature-dependent DR signal, where two distinct NIR regimes are established. One is the bulk zone, while the other is the topological regime associated with surface states. The bulk regime exhibits a variety of optically allowed transitions between valence and conduction bands. Simultaneously, the surface regime denotes the shift from a surface to a more excited state [62].

5. Temperature-dependent terahertz frequency generation

The kinetic profile is used in order to investigate the temperature dependence of the oscillation frequencies generated in the terahertz regime. In addition to being in the terahertz frequency range, other frequency modes are also observed in the TI. These modes lie in the gigahertz frequency range due to CAP. While the terahertz frequency is generated because of the vibration of COP. These modes are analyzed using filtering the high-frequency component and fitting the data (FFD) [48]. This detailed analysis helps exclude the charge carrier's relaxation from phonon-associated vibrations. The COP-associated frequency is found to have lied in the terahertz regime. Moreover, these frequency modes are dependent on excitation and probe energy, as mentioned above. In order to visualize the changes of these modes with temperature, the kinetic profile at 100 nm is considered. The micro flakes of different TI are excited

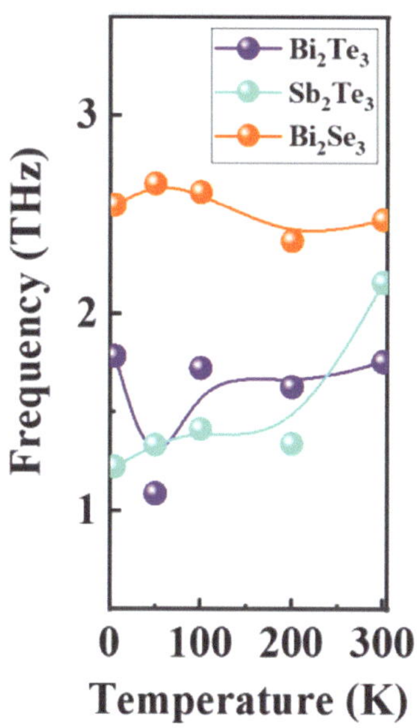

Figure 5.
Frequency dependence of three TIs on temperature. The COP modes in TIs are discovered to be temperature sensitive, implying that their vibration frequency is as well [53, 62].

with pump energy of 3.02 eV at 1.0 mW average power in order to observe the kinetic profile at this particular wavelength. The temperature-dependent investigation is carried out at these pumps and probe energy at different temperatures. With the use of this FFD analysis, it is possible to identify the terahertz frequency at which the phonons oscillate. Each of the TIs, namely Bi_2Te_3, Sb_2Te_3, and Bi_2Se_3, exhibits oscillations of the order of terahertz frequency in the range of 1–3 THz, as seen in **Figure 5**.

6. Conclusion

The low-temperature investigation convincingly demonstrates the presence of a second surface state and a wide absorbance of 1.2 eV in bismuth selenide using TRUS. While the surface state-related transition occurs at ~1.2 eV, ~1.1 eV, and ~ 1.0 eV above the Fermi surface in bismuth selenide, bismuth telluride, and antimony telluride, respectively. Thus, this work demonstrates unequivocally that temperature-dependent analysis of charge carrier and phonon dynamics aids in the extraction of surface states associated transitions within TI. Additionally, the temperature dependency of the COP mode is established in all these TIs. At 100 K, 50 K, and 5 K, the charge carrier dynamics of Bi_2Te_3 and Sb_2Te_3 show a shift from their surface state to an excited state in the conduction band, which is due to increasing carriers in the surface conduction channel. Thus, investigation of TIs at low temperatures reveals the emergence of TSS-related transitions and their dominance at low temperatures, which is repressed at room temperature due to bulk carriers' thermalization.

Acknowledgements

The director of NPL strongly supports this work. Mr. Prince Sharma wishes to express his gratitude to CSIR-UGC for financial assistance and AcSIR for admitting him as a research scholar to its Ph.D. program.

Conflict of interest

The authors declare no conflict of interest.

Author details

Prince Sharma[1,2]*, Veerpal Singh Awana[1,2] and Mahesh Kumar[1,2]

1 National Physical Laboratory (CSIR), New Delhi, India

2 Academy of Scientific and Innovative Research, Ghaziabad, Uttar Pradesh, India

*Address all correspondence to: sharmapvats8@gmail.com

References

[1] Hall EH. On a new action of the magnet on electric currents. Nature. 1880;**21**(537):361-361

[2] Von Klitzing K. The quantized Hall effect. Reviews of Modern Physics. 1986;**58**(3):519-531

[3] Hasan MZ, Kane CL. Colloquium: Topological insulators. Reviews of Modern Physics. 2010;**82**(4):3045-3067

[4] Kane CL, Mele EJ. Z2 topological order and the quantum spin hall effect. Physical Review Letters. 2005;**95**(14): 146802

[5] Ando Y, Society P, April R, Ando Y. Topological insulator materials. Journal of the Physical Society of Japan. 2013;**82**(10):1-32

[6] Moore JE. The birth of topological insulators. Nature. 2010;**464**(7286):194-198

[7] Keimer B, Moore JE. The physics of quantum materials. Nature Physics. 2017;**13**(11):1045-1055

[8] Moore J. Topological insulators: The next generation. Nature Physics. 2009;**5**(6):378-380

[9] Yan B, Zhang SC. Topological materials. Reports. Progress in Physics. 2012;**75**(9):096501

[10] Lu L, Joannopoulos JD, Soljačić M. Topological photonics. Nature Photonics. 2014;**8**(11):821-829

[11] Šmejkal L, Mokrousov Y, Yan B, MacDonald AH. Topological antiferromagnetic spintronics. Nature Physics. 2018;**14**(3):242-251

[12] Hasan MZ, Moore JE. Three-dimensional topological insulators. Annual Review of Condensed Matter Physics. 2011;**2**(1):55-78

[13] Moore JE, Balents L. Topological invariants of time-reversal-invariant band structures. Physical Review B—Condensed Matter and Materials Physics. 2007;**75**(12):3-6

[14] Qi J, Chen X, Yu W, et al. Ultrafast carrier and phonon dynamics in Bi2 Se3 crystals. Applied Physics Letters. 2010;**97**(18):1-4

[15] Kumar N, Ruzicka BA, Butch NP, et al. Spatially resolved femtosecond pump-probe study of topological insulator Bi_2Se_3. Physical Review B—Condensed Matter and Materials Physics. 2011;**83**(23):1-8

[16] Glinka YD, Babakiray S, Johnson TA, et al. Ultrafast carrier dynamics in thin films of the topological insulator Bi_2Se_3. Applied Physics Letters. 2013;**103**(15):151903

[17] Glinka YD, Babakiray S, Johnson TA, et al. Effect of carrier recombination on ultrafast carrier dynamics in thin films of the topological insulator Bi_2Se_3. Applied Physics Letters. 2014;**105**(17):171905

[18] Glinka YD, Li J, He T, et al. Clarifying ultrafast carrier dynamics in ultrathin films of the topological insulator Bi_2Se_3 using transient absorption spectroscopy. ACS Photonics. 2021;**8**(4):1191-1205

[19] Dufour D, Marchese L, Terroux M, et al. Review of terahertz technology development at INO. Journal of Infrared, Millimeter, and Terahertz Waves. 2015;**36**(10):922-946

[20] Tonouchi M. Cutting-edge terahertz technology. Nature Photonics. 2007;**1**(2):97-105

[21] Yang X, Zhao X, Yang K, et al. Biomedical applications of terahertz spectroscopy and imaging. Trends in Biotechnology. 2016;**34**(10):810-824

[22] Ferguson B, Zhang XC. Materials for terahertz science and technology. Nature Materials. 2002;**1**(1):26-33

[23] Smye SW, Chamberlain JM, Fitzgerald AJ, Berry E. The interaction between terahertz radiation and biological tissue. Physics in Medicine and Biology. 2001;**46**(9):R101-R112

[24] Crowe TW, Globus T, Woolard DL, Hesler JL. Terahertz sources and detectors and their application to biological sensing. Philosophical Transactions of the Royal Society of London. Series A: Mathematical, Physical and Engineering Sciences. 1815;**2004**(362):365-377

[25] Kampfrath T, Tanaka K, Nelson KA. Resonant and nonresonant control over matter and light by intense terahertz transients. Nature Photonics. 2013;7(9):680-690

[26] In C, Sim S, Kim B, et al. Control over electron-phonon interaction by Dirac Plasmon engineering in the Bi_2Se_3 topological insulator. Nano Letters. 2018;**18**(2):734-739

[27] Jenkins GS, Sushkov AB, Schmadel DC, et al. Terahertz Kerr and reflectivity measurements on the topological insulator Bi_2Se_3. Physical Review B: Condensed Matter and Materials Physics. 2010;**82**(12):1-9

[28] Sim S, Brahlek M, Koirala N, et al. Terahertz dynamics of topological insulator Bi_2Se_3: Ultrafast photoexcitation suppresses hot-Dirac electron surface scattering. IEEE, Conference; 2014. pp. 1-2

[29] Braun L, Mussler G, Hruban A, et al. Ultrafast photocurrents at the surface of the three-dimensional topological insulator Bi_2Se_3. Nature Communications. 2016;**7**(1):13259

[30] Kamboj VS, Singh A, Ferrus T, et al. Probing the topological surface state in Bi_2Se_3 thin films using temperature-dependent terahertz spectroscopy. ACS Photonics. 2017;**4**(11):2711-2718

[31] Valdés Aguilar R, Qi J, Brahlek M, et al. Time-resolved terahertz dynamics in thin films of the topological insulator Bi_2Se_3. Applied Physics Letters. 2015;**106**(1):011901

[32] Zhou J, Zhou T, Yang D, et al. Optically controlled extraordinary terahertz transmission of Bi_2Se_3 film modulator. Photonic Sensors. 2019;**9**(3):268-276

[33] Lee S, Sim S, Moon J, Cha S, Shin H-S, Park S et al. Ultrafast Semiconducting to Metallic Terahertz Responses in the Topological Insulator Bi_2Se_3. Optical Society of America, Conference. 2018. FF2D-4

[34] Sharma P, Kumar M, Awana VPS. Topological insulator Bi_2Se_3 as a tunable crystal for terahertz frequency generation. Applied Physics A: Materials Science & Processing. 2021;**127**(5):327

[35] Sharma P, Kumar M, Awana VPS, et al. Comprehensive analysis of terahertz frequency response of Bi_2Se_3 and Bi_2Te_3 single crystals using terahertz time-domain spectroscopy. Materials Science and Engineering B. 2021;**272**:115355

[36] Wang X, Cheng L, Zhu D, et al. Ultrafast spin-to-charge conversion at the surface of topological insulator thin films. Advanced Materials. 2018;**2018**(52):1802356

[37] Kim SHS, Shin DH, Kim JH, et al. Resonance effects in

thickness-dependent ultrafast carrier and phonon dynamics of topological insulator Bi_2Se_3. Nanotechnology. 2015;**2015**(4):045705

[38] Liu Q, Shao R, Li N, et al. Anharmonicity of Bi_2Se_3 revealed by fs transient optical spectroscopy. Applied Physics Letters. 2019;**115**(20):201902

[39] Sobota JA, Yang SL, Leuenberger D, et al. Ultrafast electron dynamics in the topological insulator Bi_2Se_3 studied by time-resolved photoemission spectroscopy. Journal of Electron Spectroscopy and Related Phenomena. 2014;**195**:249-257

[40] Sobota JA, Yang SL, Kemper AF, et al. Direct optical coupling to an unoccupied Dirac surface state in the topological insulator Bi_2Se_3. Physical Review Letters. 2013;**111**(13):136802

[41] Wang MC, Qiao S, Jiang Z, et al. Unraveling Photoinduced spin dynamics in the topological insulator Bi_2Se_3. Physical Review Letters. 2016;**116**(3):036601

[42] Cacho C, Crepaldi A, Battiato M, et al. Momentum-resolved spin dynamics of bulk and surface excited states in the topological insulator Bi_2Se_3. Physical Review Letters. 2015;**114**(9):1-6

[43] Bugini D, Boschini F, Hedayat H, et al. Ultrafast spin-polarized electron dynamics in the unoccupied topological surface state of Bi_2Se_3. Journal of Physics: Condensed Matter. 2017;**30**:30LT01

[44] Lai YP, Chen HJ, Wu KH, Liu JM. Temperature-dependent carrier-phonon coupling in topological insulator Bi_2Se_3. Applied Physics Letters. 2014;**105**(23):1-6

[45] Glinka YD, Babakiray S, Holcomb MB, Lederman D. Effect of Mn doping on ultrafast carrier dynamics in thin films of the topological insulator Bi_2Se_3. Journal of Physics. Condensed Matter. 2016;**28**(16):0-6

[46] Crepaldi A, Ressel B, Cilento F, et al. Ultrafast photodoping and effective Fermi-Dirac distribution of the Dirac particles in Bi_2Se_3. Physical Review B: Condensed Matter and Materials Physics. 2012;**86**(20):1-5

[47] Sim S, Lee S, Moon J, et al. Picosecond competing dynamics of apparent semiconducting-metallic phase transition in the topological insulator Bi_2Se_3. ACS Photonics. 2020;7(3):759-764

[48] Sharma P, Kumar M, Awana VPS. Exploration of terahertz from time-resolved ultrafast spectroscopy in single-crystal Bi_2Se_3 topological insulator. Journal of Materials Science: Materials in Electronics. 2020;**31**(10):7959-7967

[49] Hu J, Igarashi K, Sasagawa T, et al. Femtosecond study of A_{1g} phonons in the strong 3D topological insulators: From pump-probe to coherent control. Applied Physics Letters. 2018;**2018**(3):031901

[50] Jnawali G, Linser S, Shojaei IA, et al. Revealing optical transitions and carrier recombination dynamics within the bulk band structure of Bi_2Se_3. Nano Letters. 2018;**18**(9):5875-5884

[51] Flock J, Dekorsy T, Misochko OV. Coherent lattice dynamics of the topological insulator Bi_2Te_3 probed by ultrafast spectroscopy. Applied Physics Letters. 2014;**105**(1):011902

[52] Luo CW, Chen H-J, Wang HJ et al. Ultrafast dynamics in topological insulators. In Betz M, Elezzabi AY, Song J-J, Tsen K-T (eds): Ultrafast Phenom. Nanophotonics XVII, 2013; 8623:86230D.

[53] Sharma P, Sharma R, Awana VPS, et al. Low-temperature ultrafast optical probing of topological bismuth selenide.

Journal of Alloys and Compounds. 2021;**886**:161235

[54] Sharma P, Sharma MM, Kumar M, Awana VPS. Metal doping in topological insulators—A key for tunable generation of terahertz. Solid State Communications. 2020;**319**:114005

[55] Glinka YD, Babakiray S, Johnson TA, et al. Acoustic phonon dynamics in thin-films of the topological insulator Bi_2Se_3. Journal of Applied Physics. 2015;**165703**(16):1-6

[56] Gupta BK, Sultana R, Singh S, et al. Unexplored photoluminescence from bulk and mechanically exfoliated few layers of Bi_2Te_3. Scientific Reports. 2018;**8**(1):8-13

[57] Shrestha K, Graf D, Marinova V, et al. Weak antilocalization effect due to topological surface states in $Bi_2Se_{2.1}Te_{0.9}$. Journal of Applied Physics. 2017;**122**(14):145901

[58] Zhang SX, McDonald RD, Shekhter A, et al. Magneto-resistance up to 60 tesla in topological insulator Bi_2Te_3 thin films. Applied Physics Letters. 2012;**101**(20):20-24

[59] Yu X, He L, Lang M, et al. Separation of top and bottom surface conduction in Bi_2Te_3 thin films. Nanotechnology. 2013;**24**(1):6-12

[60] Assaf BA, Katmis F, Wei P, et al. Quantum coherent transport in SnTe topological crystalline insulator thin films. Applied Physics Letters. 2014;**105**(10):102108

[61] Utterback JK, Ruzicka JL, Hamby H, et al. Temperature-dependent transient absorption spectroscopy elucidates trapped-hole dynamics in CdS and CdSe Nanorods. Journal of Physical Chemistry Letters. 2019;**10**(11):2782-2787

[62] Sharma P, Kumar Y, Awana VPS, Kumar M. Temperature dependent evolution of topological surface states. Solid State Sciences. 2022;**125**:106829

Chapter 8

Recording of Micro/Nanosized Elements on Thin Films of Glassy Chalcogenide Semiconductors by Optical Radiation

V.V. Petrov, A.A. Kryuchyn, V.M. Rubish and M.L. Trunov

Abstract

Inorganic resists based on chalcogenide glassy semiconductor (CGS) films can be effectively used in the creation of micro- and nanoelements of optoelectronic devices, micro- and nanoelectromechanical systems, and diffractive optical elements. The use of these materials is based mainly on their sensitivity to different types of radiation, which causes phase and structural changes in CGS films, and transparency in the infrared range. A number of photoinduced changes are observed in CGS, which are associated with structural transformations, phase transitions, defect formation, and atomic diffusion. It is important to determine technologies for the formation of micro- and nanoscale structures on CGS films, which can be used in the creation of diffractive optical elements for optoelectronic devices. Increasing the resolution of recording media based on vitreous chalcogenide semiconductors can be achieved by choosing the recording modes and composition of glasses, in which the strongest nonlinearity of the exposure characteristics of photosensitive material, as well as the introduction into the structure of recording media nanoparticles of noble metals for excitation of plasmonic resonance.

Keywords: chalcogenide glassy semiconductor, inorganic resists, plasmonic resonance, nanoelements, near-field recording

1. Introduction

Lithography is one of the key technological processes in the production of semiconductor integrated circuits, storage devices, and precision devices for optics and micromechanics. Laser lithography, near-field, and probe technologies, and radiation exposure methods are widely used to create nanoscale structures [1, 2]. Among the methods of obtaining nanoscale structures, it is necessary to note laser lithography, which allows to carry out mask less image formation in the photoresist layer on the surface of the substrate with a laser beam. Research in the development and manufacture of nanostructures for various purposes is largely determined by the level of development of technologies that allow atomic accuracy to obtain nanostructures of the required configuration and dimension, as well as a comprehensive diagnosis of the properties of nanostructures. Modern methods of nanoscale optical recording are based on the use of various methods, including photoinitiation with high beam intensity. A significant limitation on the size of the elements is due to the ability to focus optical radiation to sizes smaller than the diffraction limit. Many technical solutions are proposed and developed to realize the possibilities of optical nanolithography [1–3]. Special photosensitive materials can be used to form nanosized structures with focused laser radiation. Direct optical lithography can be created without the use of organic photoresists of functional inorganic nanomaterials. The ability to directly pattern completely inorganic layers using a radiation dose comparable to that of organic photoresists provides an alternative method for producing thin-film devices [4, 5]. Among these materials should be noted inorganic resist based on films of chalcogenide glassy semiconductors (CVS). Inorganic resists based on CVS films can be effectively used in the creation of micro- and nanoelements of optoelectronic devices, micro- and nanoelectromechanical systems (MEMS/NEMS), and diffractive optical elements. The use of these materials is based mainly on their sensitivity to different types of radiation, which cause phase and structural changes in CVS films, and transparency in the infrared range [4, 6–9]. Numerous studies have been conducted aimed at studying the processes of formation of nanostructures on CVS films [7, 9–12]. A number of photoinduced changes are observed in CVS, which are associated with structural transformations, phase transitions, defect formation, and atomic diffusion [6–8]. It is important to determine technologies for the formation of micro- and nanoscale structures on CVS films, which can be used in the creation of diffractive optical elements for optoelectronic devices.

Photoinduced transformations in CVS have been widely used in the optical recording of information to create optical media of various types. This is largely because the properties of thin chalcogenide films allow recording at high speed and do not impose restrictions on the minimum size of the recorded elements and, consequently, on the density of information recording [7, 12]. Numerous theoretical and experimental studies of the interaction of electromagnetic radiation and particle fluxes with CVS have shown the possibility of achieving ultra-high resolution when exposed to optical radiation or electron beam [10, 11, 13, 14]. It is shown that the resolution of an inorganic resist based on CVS is determined by the size of the structural units that form the matrix of films and is 1–2 nm [10, 13, 14]. In this regard, CVS are promising materials for the formation of nanoscale structures on their surface and the creation of ultra-dense information recording devices [1]. Numerous experiments on the exposure of thin films of CVS and their subsequent selective chemical etching have shown that when using immersion optical systems, it is possible to record microrelief structures with submicron dimensions (120–150 nm). With electron beam exposure of chalcogenide films, it is possible to form structures with sizes of 50–70 nm [1, 7]. Particular attention was paid to the choice of CVS that do not contain highly toxic elements, modes of their heat treatment and selective chemical etching [15].

2. Thermolithographic recording on chalcogenide films

The diffraction nature of light prevents us from achieving sub-diffraction or nanometer resolution in an optical beam lithography system. It is necessary to develop special recording methods based on nonlinear interaction with photosensitive materials and conversion of the energy of incident radiation. The size of the elements formed on the films of CVS was determined mainly by the resolution of the optical focusing system and the accuracy of the automatic focusing system. Previous analysis showed that microrelief structures on CVS films, the width of which is less than the resolution of diffraction-limited optics, can be created using the thermolithographic recording mode. Local heating of the film in the recording area by radiation with a non-uniform intensity distribution allows to reduce the size of the elements by selecting the recording mode [1, 7, 16, 17].

The recording of elements on photosensitive materials is carried out by a focused beam with a non-uniform intensity distribution (usually with a Gaussian intensity distribution) and this allows on a photosensitive material with a nonlinear exposure characteristic to record elements smaller than the diameter of the focused beam. One of the first experiments on recording on a semiconductor–metal material with a nonlinear exposure characteristic showed the possibility of reducing the size of the elements recorded by laser radiation by 2-3 times compared to the diameter of the laser beam measured at 1/e [18] formed on the films of chalcogenide vitreous semiconductors was determined mainly by the resolution of the optical focusing system and the accuracy of the automatic focusing system. Previous analysis showed that microrelief structures on CVS films, the width of which is less than the resolution of diffraction-limited optics, can be created using the thermolithographic recording mode. The local heating of the film in the recording area by radiation with a non-uniform intensity distribution allows to reduce the size of the elements by selecting the recording mode [19–21].

The minimum size of elements recorded on films of chalcogenide semiconductors with phase transitions by thermolithography was 100 nm [22]. The use of this technology was of interest for optical disc recording systems in CD, DVD, BD [7, 16], and the creation of diffractive optical elements [1]. Examples of thermolithographic recording on chalcogenide semiconductor films are shown in **Figure 1**. The recording beam diameter was 1.0 mm. Elements 0.7-0.8 μm wide were recorded [7, 15].

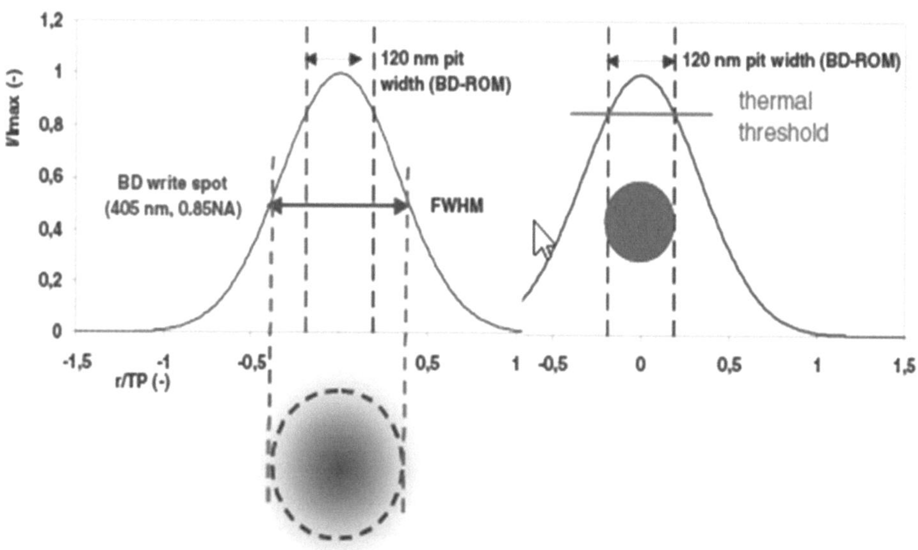

Figure 1.
The principle of the thermolithographic recording [21].

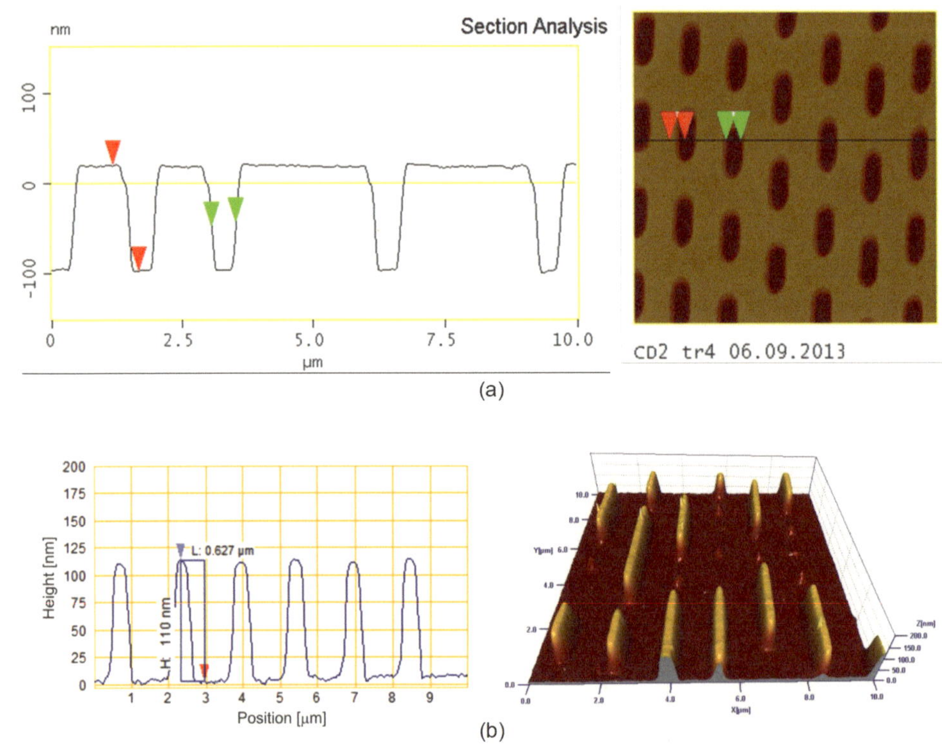

Figure 2.
(a) the relief microstructure on the positive inorganic photoresist $GeSe_2$. [15]. (b) the relief microstructure on the negative inorganic resist As_2S_3 [7].

Limitations of the thermolithographic recording mode are associated with a significant effect of fluctuations in the power of laser radiation on the size of the elements recorded on the nonlinear photosensitive material (**Figure 2**). The formation of nanoscale relief structures on thin films of CVS by diffraction-limited optical systems is problematic.

3. Use of near-field optical radiation focusing systems for the formation of submicron structures

For the formation of nanoscale elements and structures, it was proposed to use near-field systems for focusing optical radiation. Lines with a width of 100 nm and a depth of 23 nm on the As_2S_3 film were recorded with a near-field probe with an aperture of 120 nm [23]. The main disadvantage of recording by this method is the low scanning speed (100 µm/s), which is due to the low efficiency of near-field probes based on conical optical fibers [2, 23]. The situation with the use of near-field probes for recording nanoscale elements on inorganic resistors may change with the creation of new more efficient probes for focusing laser radiation, in particular, microstrip pyramidal probes [2]. Images of different types of probes are shown in **Figure 3**.

The pyramid-type microstrip probe (PTMP) has a transparent pyramid-like core with a truncated corner. Metal strips coat two opposite sidewalls of the pyramid. The transparent body and two metal strips form a tapering microstrip line, similar to an ordinary microstrip line where two opposite sides of a dielectric rectangular slab are coated with metal films, as shown in **Figure 3**. The incident beam (either a focused beam or a dielectric waveguide mode) couples to the probe through its wide end, and propagates along the probe, reaching the narrow end that forms the aperture. The light passing through the narrow end interacts with the scanned sample. In far-infrared band metal strips can be represented with high accuracy as perfect conductors which can support quasi-TEM wave which has no cut-off size. The incident light should have electric field polarization orthogonal to the metal strips in order to excite the quasi-TEM mode that has no cut-off size. A microstrip probe has a significant advantage over a conventional near-field probe in far-field transmission coefficient, especially for the small aperture size ($a < 100$ nm) since it decreases with a decrease of the aperture size as a square of the aperture diameter.

The near-field recording mode can be realized using a nanoscale diaphragm on the surface of the photosensitive material. To record nanosized elements by diffraction-limited optical systems, it was previously proposed to place an additional masking layer on the photosensitive layer, which changes the refractive index at elevated temperatures, such as a semiconductor film with a large band gap [24]. For the formation of nanosized elements on a thin film of chalcogenide semiconductor, the technology of exposure through the diaphragm (mask) was implemented, created in a material with a nonlinear exposure characteristic (technology of high-resolution near-field storage Super-RENS). One of the main elements of the Super-RENS disk is a mask, which is used to form a light beam of minimum size to expose the photosensitive layer of chalcogenide semiconductor and separated from it by a protective layer of fixed thickness. The size of the optical spot created by the mask ultimately determines the size of the elements that are recorded in the media. The scheme of the Super-RENS recording method is shown in **Figure 4b** [25].

As the material of the diaphragm in the first experiments, thin antimony (Sb) films with a thickness of 15 nm were used, located between the protective layers of SiN (**Figure 4**) [25]. Significant optical nonlinearity of the thin antimony film located between the dielectric layers was detected. The change in transmittance was

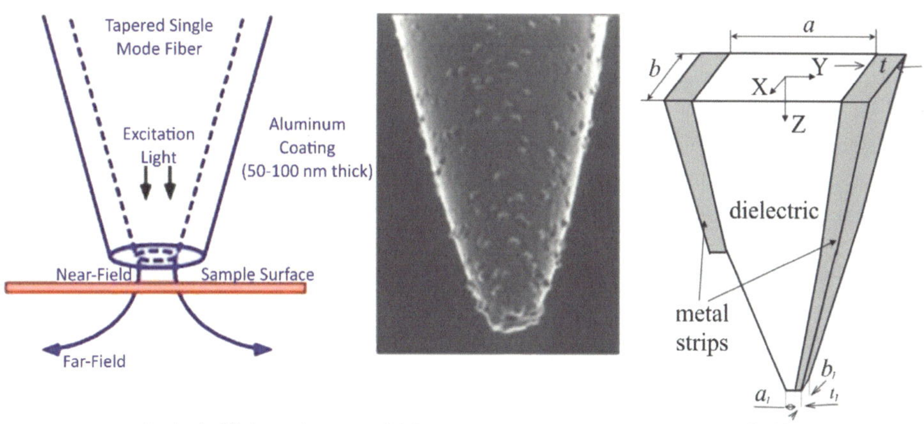

Optical efficiency is 10^{-6} at $\lambda/d = 0.1$, where d is aperture diameter
(a)

Optical efficiency is $2 \cdot 10^{-1}$ at $\lambda/d = 0.1$
(b)

Figure 3.
(a) Most commonly used near-field tapered fiber probe, (b) near-field probe based on optical plasmon microstrip line (optical microstrip probe) [2].

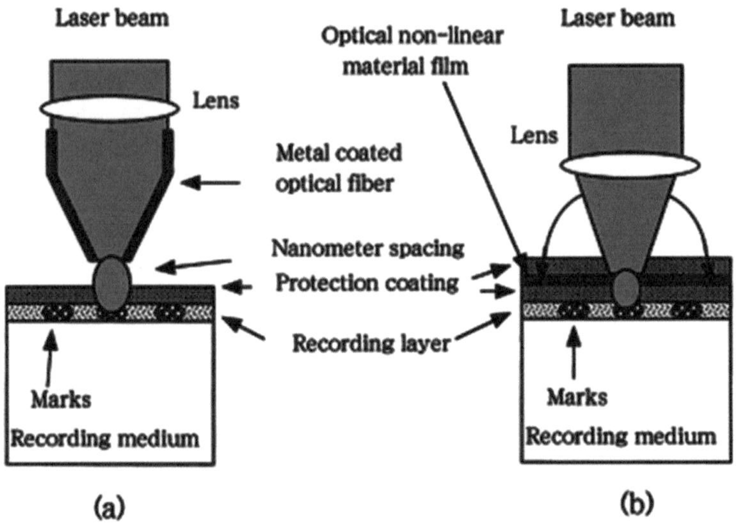

Figure 4.
(a) Conventional recording by NSOM, (b) super-RENS recording method [25].

relatively significant and stable over time in the region with submicron dimensions. This recording technology made it possible to obtain fingerprints with linear dimensions of less than 100 nm on GeSbTe (GST) films [25]. In this structure, a thin film mask made of Sb was placed at a distance of the near field to the recording layer. It was found that the Sb_2Te_3 material can also be used as a masking layer. Using it, fingerprints with linear dimensions of 60 nm were recorded on a phase-transition material (GeSbTe). The dielectric material ZnS-SiO2 was used as a protective layer. The possibility of recording and reproducing elements with submicron dimensions is explained by the fact that in the process of recording and reading in the masking layer a small hole is formed, which functions as a local solid-state near-field lens [26].

4. The use of nanocomposite photosensitive materials, the formation of images on which is carried out using the methods of nanoplasmonics

Plasmonic nanolithography, which uses surface plasmons to create submicron elements, is a promising technology for producing nanoscale structures. Plasmonics can focus light into zones smaller than the diffraction limit, due to the connection of light with the surface collective vibrations of free electrons at the metal-dielectric interface. Surface plasmon resonances (SPP) have been used to create nanoscale structures [27]. The method of plasmon nanolithography is being developed in which metal lattice masks are used to excite SPP and structural nanoscale elements. The mask is in close contact with the photoresist applied to the substrate. Typically, the incident light passes through the mask through the SPP and is directed to the photoresist [27]. It was demonstrated that the use of surface plasmons in the optical near field of a metallic mask can produce fine patterns with a subwavelength resolution. Using a silver grating mask with 300 nm periodicity, lithography with 100 nm pitch has been demonstrated by using the interference of surface plasmon waves within the grating area [28].

The method of plasmon nanolithography was used for the alternative design of the Super-RENS recording method. In this method, a layer of noble metal oxides (AgO_x, PtO_x, and PdO_x) was used instead of the Sb layer. Using surface plasmons has greater possibilities for the creation of super-dense recording systems. Irradiation of the oxide layer led to the decomposition of the oxide and the formation of a layer of metal nanoparticles. The process of chemical decomposition occurs in the temperature range from 400°C to ~500°C. Surface plasmons, excited by light on the formed nanoparticles of precious metals, generate optical near-field radiation, which is exposed to the photosensitive layer. The structure of such a medium is shown in **Figure 5** [26]. The media with the Ag_2O layer were studied in the most detail. The Ag_2O layer in the Super-RENS carrier acts as a center of strong light scattering in the local region of the multilayer carrier. The optical near field, which is created around the scattering center with Ag_2O, is 40 times stronger than the field created by the antimony layer [29, 30]. Studies have shown that the higher efficiency of high-resolution super-RENS disks with an AgO_x layer is associated with the formation of localized surface plasmons by silver clusters dissociated from the AgO_x layer. The diameter of the silver nanoparticles was approximately 4 nm. The density and distribution of dissociated silver nanoparticles are affected by the intensity of focused laser radiation. Localized surface plasmons improve the reading efficiency in such media [31].

One of the possible ways to overcome the diffraction barrier can be the use of the near light field of metal nanoparticles (NPs) integrated into chalcogenide films, i.e., the formation of a kind of plasmonic nanostructures [5]. This field arises upon irradiation with light with a certain wavelength due to the excitation of collective oscillations of free electrons in NPs (surface plasmon resonance (SPR)). The spatial distribution of this field can be changed in a controlled manner due to appropriate changes in the size and geometry of the woofer. The technology of excitation of metal nanoparticles and the use of optical near-field radiation for the exposure of photosensitive layers has proved to be quite effective and continues to develop in the creation of new types of media for recording nanoscale structures. A schematic representation of the information carrier with a layer of nanoparticles of precious metals is shown in **Figure 6** [32].

Nanoparticles of precious metals with sizes of the order of tens of nanometers can have a significant impact on the processes of recording information in different types of optical and magnetic media. The technology of using nanoparticles is one of the ways to overcome the diffraction limit in the process of recording nano-sized structures. The generation of localized plasmons in noble metal nanoparticles is widely

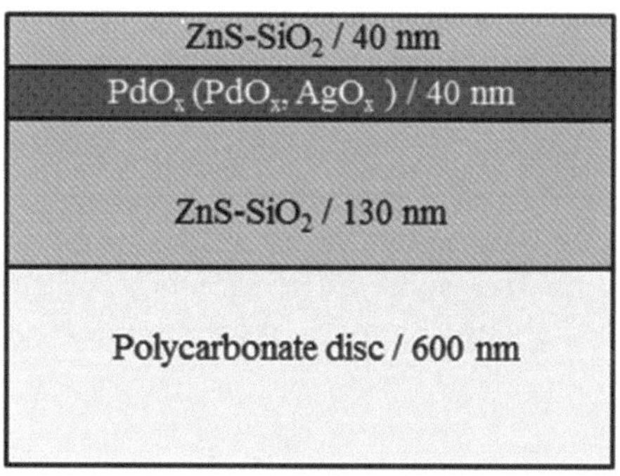

Figure 5.
The structure of the carrier made by technology super-RENS using oxides of noble metals [26].

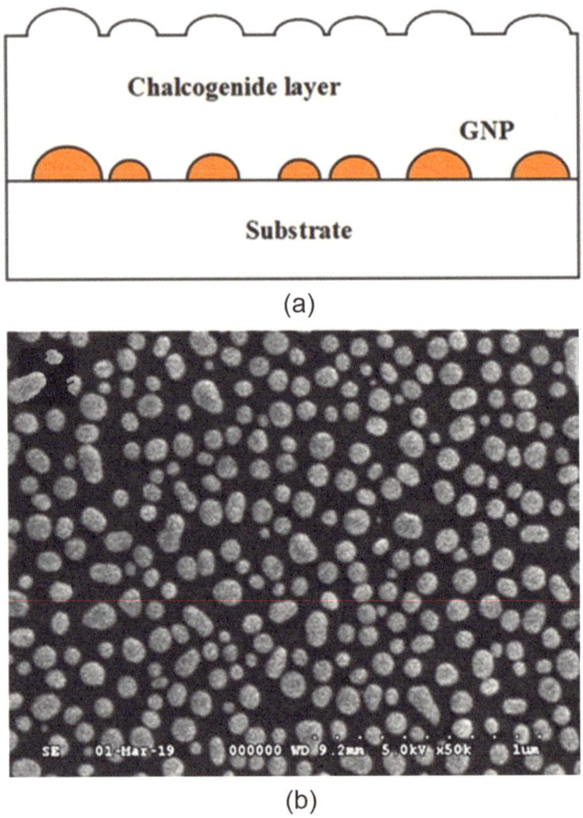

Figure 6.
Information carrier with a layer of noble metal nanoparticles: (a) schematic diagram (cross-section) of the created structures with gold nanoparticles (GNPs) and chalcogenide layer; (b) SEM picture of the created GNPs [32].

used to enhance the interaction of light with the matrix surrounding these plasmon nanostructures. The incident light, which is absorbed by the nanoparticles and transforms into collective oscillations of free electrons in them, leads to a strong amplification of the local electric field [31]. Metal nanoparticles effectively absorb light. Their ability to focus light in small volumes has led to the use of woofer in a variety of areas, including as light concentrators for the solar cells. The light-concentrating properties of metal nanostructures are a consequence of the amplification of electromagnetic fields due to the generation of localized plasmons [31, 33–35]. Light-induced plasmon heating of a magnetic medium in the process of magnetic recording (with a built-in plasmon antenna) can be used to implement the mode of thermal assistance and, ultimately, to increase the density of information recording [31, 33]. The surface plasmon interference nanolithography (SPIN) allows to obtained uniform interference patterns far beyond the free-space diffraction limit of the light. This technique provides a new alternative fabrication method for nanodevices [28].

5. Formation of microrelief structures on the surface of chalcogenide films

One of the most promising uses of optical and electron beam recording on CVS is a direct one-step process of microrelief formation on the surface of films, which is closely related to induced mass transfer (vertical or lateral directions) in amorphous material under illumination [4, 36–39]. The formation of relief on the surface of ChS films is possible by a purely optical method due to photoinduced mass transfer (FM) under the action of the light from the spectral region of the absorption edge even at relatively low intensities of the light wave. However, due to light diffraction, the lateral scale of such topographic structures is limited, which blocks the formation of nanoscale information elements [3]. In direct one-stage laser or electron-beam recording, there is an irreversible amplitude-phase optical and geometric structuring of the surface. This effect can be used for the manufacture of microlenses, amplitude-phase optical elements. The process of direct photoinduced fabrication of microrelief structures on CVS films by lateral mass transfer was studied on films of different compositions and with different irradiation methods. As a result, the observed process models were proposed and areas of possible application were identified [4, 36–38]. The possibility of creating planar diffraction optical elements during electron-beam exposure with a local change in the refractive index was experimentally demonstrated. The lens is created in the form of electron beam-recorded annular zones with a stepwise decrease in refractive index. The image of the Fresnel lens obtained by this method is shown in **Figure 7**. The minimum width of the elements in the recorded image is ~0.6–1.0 μm [31, 39].

Direct single-stage laser or electron beam recording is more efficient in nanosized layered structures Se/As_2S_3 and Sb/As_2S_3 than in homogeneous layers of As_2S_3 [6, 39]. The effect of photoinduced mass transfer allows to obtain holographic gratings, integral optical elements by a purely optical method at relatively low intensities of light fluxes [4, 13, 38, 39]. The image of the diffraction grating obtained by the direct (optical) method due to photoinduced mass transfer of the substance of the CVS film is shown in **Figure 8**.

The relief shape and diffraction efficiency can be changed by the ratio of the polarization of the recording rays and the beam of additional illumination.

The photoinduced changes in amorphous Ge-based chalcogenide layers deposited on gold nanoparticles change significantly. The rate and final magnitude of the volume change is higher in a structure with localized plasmon fields, mainly because the latter affects the processes of charge generation and the movement of atoms, initiated by illumination. The results showed that the superposition of the localized plasmon field of nanoparticles with the electromagnetic field of incident photons during irradiation enhance light-induced transformations (**Figure 9**) [32].

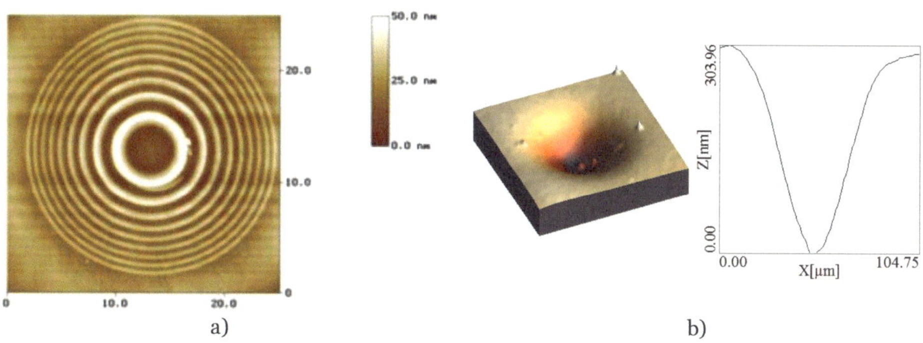

Figure 7.
Image of a Fresnel lens obtained by the method of direct one-stage process forming microrelief (a), e-beam fabricated scattering lens on Se/As2S3 NML and its profile measured by AFM (b) [6].

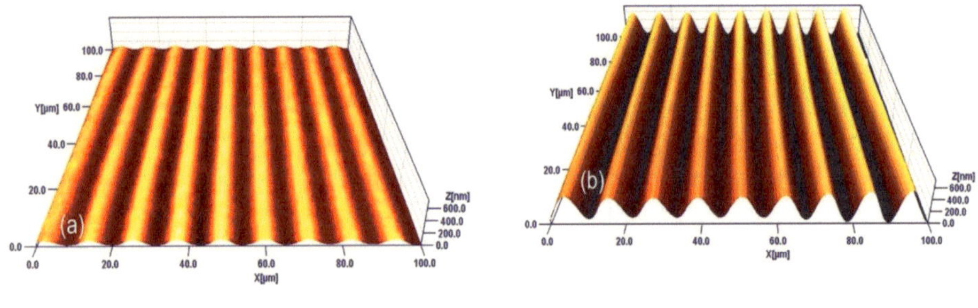

Figure 8.
AFM image of diffraction gratings recorded at temperatures of 77 K (a) and 300 K (b) on $As_{20}Se_{80}$ thin films [38].

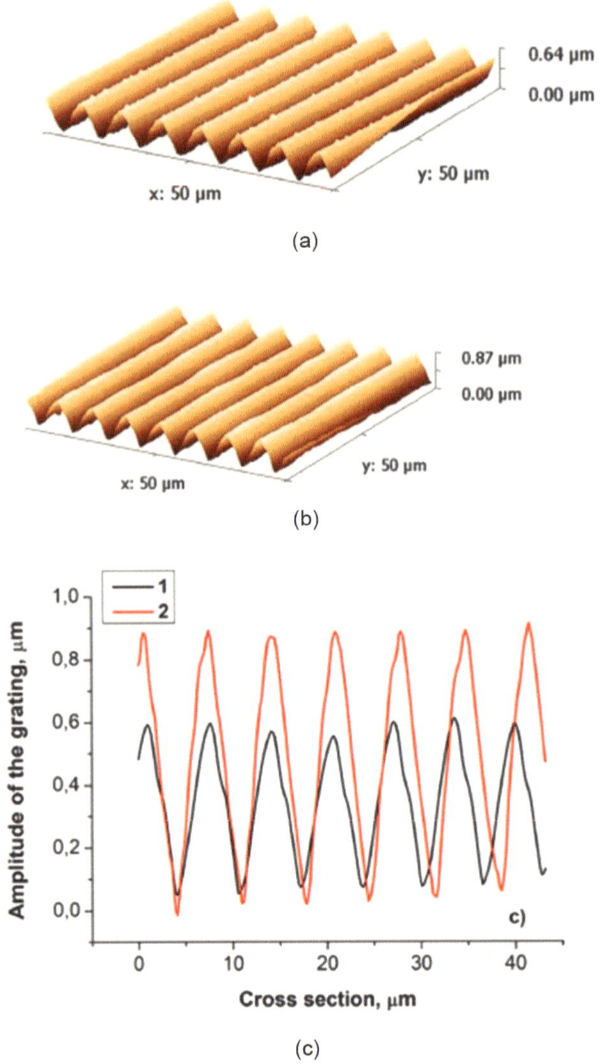

Figure 9.
AFM surface morphology of the holographic grating recorded in the pure chalcogenide layer (a) and in the sample with GNPs (b), and the cross-section of the created surface structures (c): 1—pure chalcogenide layer, 2—sample with GNP [32].

The use of a layer of gold nanoparticles allows a higher level of mass transfer and more efficient modification of the surface of a chalcogenide semiconductor film to be realized.

6. The formation of nanosized relief structures on films of chalcogenide vitreous semiconductors by the methods of probe microscopy

Scanning probe lithography (SPL) is a direct-write nanolithography technique in which elements on thin films of chalcogenide semiconductors are created by scanning a sample with a sharp nanometer tip to create localized modifications. The interactions of the tip with the sample are varied and can include mechanical, electrical, diffusion, and thermal effects. SPL techniques are being studied intensively. SPL methods provide nanometer resolution, however, they are characterized by a low write speed, which is 0.1–50 μm/s [40]. The technology of nano-heating of thin-film materials with a phase transition has been studied in detail and it has been shown that it makes it possible to register ultra-small spots with sizes less than 50 nm [41].

With the use of technologies based on atomic force microscopy, the recording of data and the formation of nanosized structures on films of CVS with phase transitions were successfully executed [42]. The information recording is based on the formation of local (crystalline/amorphous) sections with different structures and, respectively, conductivities in the nanosized layer of a chalcogenide material with the help of electric pulses [42]. It was shown that, due to the applied pulses of the voltage between a probe and a conducting electrode, the conductance of a chalcogenide amorphous $GeSb_2Te_4$ film increases by at least two orders. An increase in the conductance is caused by the phase transition of a chalcogenide film from the amorphous state into a crystalline one. The recorded data are read with the probe of an AFM by the measurement of changes in the conductance of a chalcogenide film. The simultaneous measurement of the conductance and topographic images with the help of an AFM showed that the surface relief of recorded zones is invariable in the process of recording. The least recorded imprints were down to 10 nm in diameter [42].

Such technology of recording allows one to form elements with sizes of a recording zone of ~30-70 nm, which is 3-5 times less than the real diameter of focused exposing rays of laser sources of violet and ultraviolet emissions.

The creation of a relevant relief in a recording zone occurs directly during the simultaneous exposure and indentation, which presents the essential advantage over the available lithographic methods, which require the additional treatment of a carrier with selective chemical etchants or the ion-beam or plasma chemical etching. The profiles of imprints obtained at the nanoindentation of an $As_{20}Se_{80}$ film in dark and light at various loads are given in **Figure 10** [1].

One of the advantages of the given method is the absence of a thermal heating source used for the softening of the surface area that is modified and, respectively, the absence of shortcomings related to such heating (power losses, the complexity of a micromechanical system, loads, etc).

An extension of scanning probe lithography (SPL) is plasmonic nanolithography with a focused beam. This technology is promising due to its sub-diffraction resolution. In this method, the resist is scanned and illuminated by a focused light spot created by a plasmonic lens.

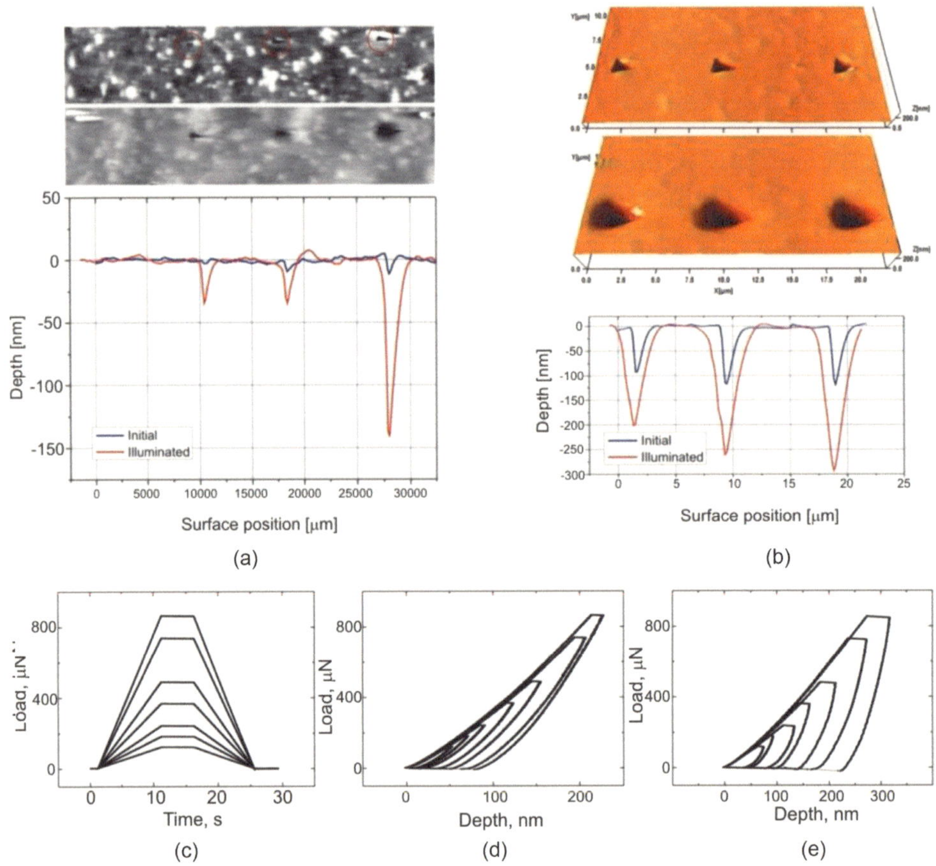

Figure 10.
General view of imprints (upper row) and their profiles (lower row) obtained at the nanoindentation of an $As_{20}Se_{80}$ film in dark and in light at loads of 120-240 mN (a) and 700-900 mN (b). Scheme of application of a load (c) and the relevant curves of nanoindentation in dark (d) and in light (e) [1, 43].

7. Conclusions

1. The physicochemical properties of glassy chalcogenide semiconductors make it possible to record nanosized elements on them under the action of actinic radiation. When recording nanoscale elements, the main problem is to focus the radiation to the required size. Increasing the resolution of recording media based on vitreous chalcogenide semiconductors can be achieved by choosing recording modes and glass composition, in which the strongest nonlinearity of the exposure characteristics of photosensitive material.

2. Among the optical methods of formation of nano-sized structures, the method, in which the near light field of nanoparticles of noble metals integrated within a thin film of CVS is used, is of high meaning. The main limitations in the use of near-field recording methods are associated with the low recording speed of nanosized elements and the need to maintain the distance between the focusing element and the photosensitive material with high accuracy.

3. The unique properties of CVS allow the formation of microrelief images without selective chemical etching due to the irradiation of thin films with rays of different polarization. One of the main tasks when using this method is the choice of recording modes, which ensures a high rate of formation of microrelief structures.

4. As a promising method of formation of nanosized structures in films of CVSs, may be utilize the optic mechanical method based on the photo plastic effect. The formation of nanosized elements (30-70 nm) in the recording zone occurs during the simultaneous exposure and indentation of the surface of a film.

Author details

V.V. Petrov, A.A. Kryuchyn*, V.M. Rubish and M.L. Trunov
Institute for Information Recording of the National Academy of Sciences of Ukraine, Kiev, Ukraine

*Address all correspondence to: kryuchyn@gmail.com

References

[1] Kryuchyn AA, Petrov VV, Rubish VM, Trunov ML, Lytvyn PM, Kostyukevich SA. Formation of nanoscale structures on chalcogenide films. Physica Status Solidi (B). 2017;**255**:1700405. DOI: 10.1002/pssb.201700405

[2] Petrov VV, Kryuchyn AA, Kunitskii YA, Rubish VM, Lapchuk AS, Kostyukevych SA. Methods of Nanolithography. Kyiv: Naukova Dumka; 2015. p. 262

[3] Trunov ML, Lytvyn PM, Sterligov VA, Lofay F, Prokopenko IV. Formation of nanoscale structures upon photoexcitation of surface plasmon resonance in nanocomposites based on textured films of gold and chalcogenide glass // theory and experiment. Chemistry. 2018;**54**(2):98-103

[4] Achimova E. Direct surface relief formation in nanomultilayers based on chalcogenide glasses: A review. Surface Engineering and Applied Electrochemistry. 2016;**52**(5):456-468. DOI: 10.3103/S1068375516050021

[5] Wang Y, Fedin I, Zhang H, Talapin DV. (2017) Direct optical lithography of functional inorganic nanomaterials. Science. V. 357, N 6349. P. 385-388. DOI: 10.1126/science.aan2958

[6] Takats V, Miller F, Jain H, Cserhati C, Kokenyesi S. Direct surface patterning of homogeneous and nanostructured chalcogenide layers. Physica Status Solidi (C). 2009;**6**(S1):83-85. DOI: 10.1002/pssc.200881347

[7] Petrov VV, Kryuchyn AA, Rubish VM, Kostyukevich SO. Inorganic Photolithography. Kiev: IMF NASU; 2007. p. 196

[8] Isbi Y, Sternklar S, Granot E, Lyubin V, Klebanov M, Lewis A. Sub-wavelength optical recording on chalcogenide glassy film. Optics Communications. 1999;**171**(4/6):219-223. DOI: 10.1016/S0030-4018(99)00541-6

[9] Tanaka K. Photoinduced structural changes in amorphous semiconductors. Semiconductors. 1998;**32**:861-866. DOI: 10.1134/1.1187473

[10] Kovalskiy A, Vlcek M, Waits CM, Dubey M. Chalcogenide glass e-beam and photoresists for ultrathin grayscale patterning. Journal of Micro/Nanolithography, MEMS, and MOEMS. 2009;**8**(4):043012

[11] Kolobov AV, Tominaga J. Chalcogenide glasses in optical recording: Recent progress. Journal of Optoelectronics and Advanced Materials. 2002;**4**(3):679-676

[12] Petrov VV, Kryuchyn AA, Shanoilo SM, Kravets VG, Belyak YV. Super-Dense Optical Information Recording. Kyiv: Institute for Information Recording NAS of Ukraine; 2009. p. 320

[13] Kokenyesi S, Ivan I, Takats V, Pálinkás J, Biri S, Szabó I. Formation of surface structures on amorphous chalcogenide films. Journal of Non-Crystalline Solids. 2007;**353**:1470-1473. DOI: 0.1016/j.jnoncrysol.2006.09.064

[14] Vlcek M, Jain H. Nanostructuring of chalcogenide glasses using electron lithography. Journal of Optoelectronics and Advanced Materials. 2006;**8**(6):2108-2111

[15] Indutny I, Kryuchin A, Borodin Y, Danko VA, Lukaniuk MV, Minko VI,

et al. Optical recording of micro- and nano- relief structures on inorganic resists Ge-Se. Data Recording, Storage and Processing. 2013;**15**(4):3-14

[16] Kryuchyn AA, Petrov VV, Kostyukevych SO. High density optical recording in thin chalcogenide films. Journal of Optoelectronics and Advanced Materials. 2011;**13**(11-12):1487-1492

[17] Helseth LE. Breaking the diffraction limit in nonlinear materials. Optics Communications. 2005;**256**(4-6):435-438. DOI: 10.1016/j.optcom.2005.06.074

[18] Kryuchyn AA, Petrov VV. Influence of nonlinearity of the recording medium on the density of information recording in optical storage devices. Kvantovaia Elektronika. 1977;**4**(1):188-190

[19] Petrov VV, Kryuchyn AA. On the photosensitivity of systems based on glassy chalcogenide semiconductors at high irradiation power. Kvantovaia Elektronika. 1974;**1**(12):2618-2620

[20] Kostyshin MT, Kostyukevich SA. Behavior of the Ag-As_2S_3 photosensitive system under irradiation with high-power pulsed laser radiation. Ukraine Physical Journal. 1981;**26**(9):1561-1563

[21] Meinders ER, Rastogi R, Van der Veer M, Peeters P, Majdoubi HE, Bullem H, et al. Phase-transition mastering of high-density optical media. Japanese Journal of Applied Physics. 2007;**46**(6B):3987-3992

[22] Wei J, Jiao X. Super-resolution nanopatterns and optical recording in chalcogenide phase change thin films by direct laser writing. In: Proceedings SPIE 7125, Eighth International Symposium on Optical Storage and 2008 International Workshop on Information Data Storage, 712505. 2009. DOI: 10.1117/12.825533

[23] Noach S, Manevich M, Eisenberg NP, Davidov D, Klebanov M, Lubin V. Optical near-field lithography in halcogenidefilms. Optical Materials. 2006;**28**:1054-1057. DOI: 10.1016/j.optmat.2005.06.004

[24] Wu Y, Chong CT. Theoretical analysis of a thermally induced superresolution optical disk with different readout optics. Applied Optics. 1997;**36**:6668-6682. DOI: 10.1364/AO.36.006668

[25] Tominaga J, Nakano T, Atoda N. An approach for recording and readout beyond the diffraction limit with an Sb thin film. Applied Physics Letters. 1998;**73**:2078. DOI: 10.1063/1.122383

[26] Shi LP, Chong TC, Yao HB, Tan PK, Miao XS. Super-resolution near-field optical disk with an additional localized surface plasmon-coupling layer. Journal of Applied Physics. 2002;**91**:10209. DOI: 10.1063/1.1476068

[27] Chorsi H. (2020). Plasmonics as a fabrication tool. 2020. Available from: https://arxiv.org/abs/2005.06005v

[28] Liu ZW, Wei QH, Zhang X. Surface plasmon interference nanolithography. Nano Letters. 2005;**5**(5):957-961. DOI: 10.1021/nl0506094

[29] Liu Q, Fukaya T, Tominaga J, Iwanabe Y, Shima T. Optical properties of metal-oxide films in super-RENS. Japanese Journal of Applied Physics. 2005;**44**(7A):5156-5163. DOI: 10.1143/JJAP.44.5156

[30] Fukaya T, Buchel D, Shinbori S, Tominaga J, Atoda N. Micro-optical nonlinearity of a silver oxide layer. Journal of Applied Physics. 2001;**89**:6139. DOI: 10.1063/1.1365434

[31] Chu TC, Liu W-C, Tsai DP. Enhanced resolution induced by random silver

nanoparticles in near-field optical disks. Optics Communications. 2005;**246**:561-567. DOI: 10.1016/j.optcom.2004.11.001

[32] Csarnovics I, Veres M, Nemec P, Molnár S, Kökényesi S. Surface plasmon enhanced light-induced changes in Ge-Se amorphous chalcogenide – Gold nanostructures. Journal of Non-Crystalline Solids. 2021;**553**:120491. DOI: 10.1016/j.jnoncrysol.2020.120491

[33] Kravets VG, Kabashin AV, Barnes WL, Grigorenko AN. Plasmonic surface lattice resonances: A review of properties and applications. Chemical Reviews. 2018;**118**(12):5912-5951. DOI: 10.1021%2Facs.chemrev.8b00243

[34] Trunov ML, Lytvyn PM, Nagy PM, Csik A, Rubish VM, Kokenyesi S. Light-induced mass transport in amorphous chalcogenides: Toward surface plasmon-assisted nanolithography and near-field nanoimaging. Physica Status Solidi (b). 2014;**251**(7):1354-1362. DOI: 10.1002/pssb.201350296

[35] Hartland GV, Besteiro LV, Johns P, Govorov AO. What's so hot about electrons in metal nanoparticles. ACS Energy Letters. 2017;**2**:1641-1653. DOI: 10.1021/acsenergylett. 7b00333

[36] Trunov ML, Lytvyn PM, Nagy PM, Dyachyns'ka OM. Real-time atomic force microscopy imaging of photoinduced surface deformation in AsxSe100−xAsxSe100−x chalcogenide films. Applied Physics Letters. 2010;**96**(11):111908. DOI: 10.1063/1.3360229

[37] Yannopoulos SN, Trunov ML. Photoplastic effects in chalcogenide glasses: A review. Physica Status Solidi (b). 2009;**246**(8):1773-1785. DOI: 10.1002/pssb.200982005

[38] Takáts V, Trunov ML, Vad K. Low-temperature photo-induced mass transfer in thin $As_{20}Se_{80}$ amorphous films. Materials Letters. 2015;**160**:558-561

[39] Gil D, Menon R, Smith HI. Fabrication of high numerical aperture phase zone plates with a single lithography exposure and no etching. Journal of Vacuum Science & Technology B. 2003;**21**:2956. DOI: 10.1116/1.1619957

[40] Howell ST, Grushina A, Holzner F, Brugger J. Thermal scanning probe lithography—A review. Microsystems & Nanoengineering. 2020;**6**(1):1-24. DOI: 10.1038/s41378-019-0124-8

[41] Hamann H, O'Boyle M, Martin YC, Rooks M, Wickramasinghe HK. Ultra-high-density phase-change storage and memory. Nature Materials. 2006;**5**:383-387. DOI: 10.1038/nmat1627

[42] Kado H, Tohda T. Nanometer-scale recording on chalcogenide films with an atomic force microscope. Applied Physics Letters. 1995;**66**:2961. DOI: 10.1063/1.114243

[43] Trunov ML. Photoplastic effect in non-crystalline materials: A nanoindentation study. Journal of Physics D, Applied Physics. 2008;**41**(7):9. DOI: 101088/0022-3727/41/7/074011